Regulation of Carcinogenesis, Angiogenesis and Metastasis by the Proprotein Convertases (PCs)

Regulation of Carcinogenesis, Angiogenesis and Metastasis by the Proprotein Convertases (PCs)

A New Potential Strategy in Cancer Therapy

Edited by

A-Majid Khatib
INSERM, Paris, France

 Springer

A C.I.P. Catalogue record for this book is available from the Library of Congress.

ISBN-10 1-4020-4793-2 (HB)
ISBN-13 978-1-4020-4793-0 (HB)
ISBN-10 1-4020-5132-8 (e-book)
ISBN-13 978-1-4020-5132-6 (e-book)

Published by Springer,
P.O. Box 17, 3300 AA Dordrecht, The Netherlands.

www.springer.com

Cover Legend: Cascade events implicating the proprotein convertases (PCs) in tumor growth
and metastasis. PCs control tumor cell adhesion by activating or/and inducing the expression
of adhesion molecules, regulate cell proliferation by activating growth factors, cytokines,
and their receptors and induce migration and invasion of tumor cells that leads to metastases
formation by activating MMPs.

Printed on acid-free paper

TABLE OF CONTENTS

PREFACE

To date, cancer therapies are mainly based on the use of cytotoxic drugs and/or radiotherapy that have a relatively low therapeutic index with significant side effects, although they have a potent antitumor activity, since they target both cancer and normal cells. Furthermore, several cancer types are intrinsically not sensitive or become resistant to treatment with cytotoxic drugs or radiotherapy. For this reason, there is a need for the development of novel effective anticancer agents that are more specific for cancer or less toxic for normal cells. The recent discovery regarding the implication of the proprotein convertases in the processes of tumor progression and metastasis, made these enzymes potential new targets in cancer therapy. The first studies demonstrating experimentally the importance of the convertases in cancer was reported in the year 2001. Since this period, the growing body of knowledge and evidence regarding the importance of these molecules in the activation and the regulation of molecules involved in tumor progression and metastasis such as growth factors, adhesion molecules and MMPs is now leading to the increased understanding of the processes of tumorigenesis, angiogenesis and metastasis processes.

However, although the level of the PCs was reported to be higher in various cancers and the idea to inhibit their activity and/or expression may constitute a promising new strategy in cancer therapy, these enzymes are also required for the homeostasis of normal cells. Thereby, the development of PCs inhibitors-derived drugs whose actions can target specific tissues and cell types is required. Similarly, although the use of general PC inhibitors may be advantageous, in some cases it may be necessary to target only one member of the PC family. Therefore, one of the important future developments would be to find and express PC inhibitors specific for each member of the PC family. This is feasible, as was demonstrated by the identification of the specific and natural inhibitor of the convertase PC1 (pro-SAAS) and the convertase PC2 (7B2). In the long term, these inhibitors may provide a rationale for testing this family of compounds as anti-metastatic agents or in conjunction with standard therapy in clinical settings.

In this book, leading experts and pioneers in the area of convertases research dealing with cancer provide an overview that summarizes the current state of knowledge on the role of these enzymes in carcinogenesis and metastasis.

Dr A-Majid Khatib
INSERM, Paris, France

LIST OF CONTRIBUTORS

Upendra K. Banik.
Bioscan Continental Inc., 350, Industriel Boul., St-Eustache, Quebec J7R 5R4, Canada

Ajoy Basak.
Diseases of Aging Program, Regional Protein Chemistry Center, Ottawa Health Research Institute, Loeb Building, 725 Parkdale Ave, Ottawa, ON K1Y 4K9

Daniel E. Bassi.
Department of Pathology, Fox Chase Cancer Center, Philadelphia, Pennsylvania 19111, USA

Fabien Calvo.
INSERM U 716/AVENIR, IGM, Rue Juliette Dodu, 75010, Paris, France

Luis J.
CNRS FRE2737, Faculté de Pharmacie, 27 Bd J. Moulin, 13 385 Marseille Cedex 5, France

Michel Chrétien.
Regional Protein Chemistry Centre, Diseases of Ageing Unit, OHRI, Loeb Building, 725 Parkdale Ave., Ottawa, Ontario, Canada K1Y 4E9

Claire M. Dubois.
Immunology Division, Faculty of Medicine, Université de Sherbrooke, Sherbrooke, Qc, Canada, J1H 5N4

Majid Khatib.
INSERM U 716/AVENIR, IGM, 27, Ruc Juliettc Dodu, 75010, Paris, France

Andrés J.P. Klein-Szanto.
Department of Pathology, Fox Chase Cancer Center, Philadelphia, Pennsylvania 19111, USA

Stephanie McMahon.
Immunology Division, Faculty of Medicine, Université de Sherbrooke, Sherbrooke, Qc, Canada, J1H 5N4

Shawn K. Murray.
Department of Pathology, Tupper Medical Building, 5850 College Street, Dalhousie UniversityHalifax, NS, Canada, B3H 1X5

Mark W. Nachtigal.
Department of Pharmacology, Department of Pathology, Tupper Medical Building, 5850 College Street, Dalhousie UniversityHalifax, NS, Canada, B3H 1X5

Rigot V.
CNRS FRE2737, Faculté de Pharmacie, 27 Bd J. Moulin, 13 385 Marseille Cedex 5, France

Nathalie Scamuffa.
INSERM U 716/AVENIR, IGM, 27, Rue Juliette Dodu, 75010, Paris, France

Nabil G. Seidah.
Laboratory of Biochemical Neuroendocrinology, Institut de Recherches Cliniques de Montréal, 110 pine Ave West, Montreal, Quebec, Canada H2W 1R7

Geraldine Siegfried.
INSERM U.143, Kremlin-Bicetre, France

Sarmistha Basak.
Diseases of Aging Program, Regional Protein Chemistry Center, Ottawa Health Research Institute, Loeb Building, 725 Parkdale Ave, Ottawa, ON K1Y 4K9

Suiyang Li.
Bioscan Continental Inc., 350, Industriel Boul., St-Eustache, Quebec J7R 5R4, Canada

Brigitte L. Thériault.
Department of Pharmacology, Tupper Medical Building, 5850 College Street, Dalhousie UniversityHalifax, NS, Canada, B3H 1X5

Yangxin Fu.
Department of Pharmacology, Tupper Medical Building, 5850 College Street, Dalhousie UniversityHalifax, NS, Canada, B3H 1X5

INTRODUCTION

THE EVER EXPANDING SAGA OF THE PROPROTEIN CONVERTASES: FROM BENCH TO BEDSIDE

NABIL G. SEIDAH

Laboratory of Biochemical Neuroendocrinology, Institut de Recherches Cliniques de Montreal, 110 pine Ave West, Montreal, Quebec, Canada H2W 1R7

The number of protein/peptide products that result from a given genome depends on multiple factors that generate both diversity and specificity. Prominent among these are processes that regulate post-translational modifications of the primary product of mRNA translation, the precursor protein. The primary events governing the modification of the amino acid chain of secretory proteins include the N-glycosylation and signal peptide cleavage by signal peptide peptidase. This is then followed by trimming of the glycosylation chain and remodeling up until it reaches its final form in the Trans Golgi Network (TGN). Since the early/mid 1960s it was realized that most secretory proteins undergo at least one peptide bond cleavage along their trafficking pathway, e.g., by signal peptide peptidase in the endoplasmic reticulum (ER) and/or by one or more proteinase in the Golgi apparatus to release the final form of the protein and/or its processing products. Proteolysis is essentially an irreversible process, because no known enzyme can repair broken peptide bonds under normal physiological conditions. The primary event of peptide bond cleavage induces conformational changes in the resulting product, thereby generating productive biological activity. The repertoire of the secretory protein precursors that undergo limited proteolysis is large and varied. It includes many proteins that are translocated across membranes such as polypeptide endocrine and neural hormones, growth factors and their receptors, membrane bound transcription factors, adhesion molecules, extracellular matrix proteins, proteases and other types of enzymes, as well as a number of surface glycoproteins of opportunistic pathogenic viruses and bacteria.

While it is predicted that the mammalian genome codes for 460 human and 525 mouse functional proteases [1], only a handful of these are implicated in the intracellular limited proteolysis of precursor proteins.

1

A-Majid Khatib (ed.), Regulation of Carcinogenesis, Angiogenesis and Metastasis by the Proprotein Convertases, 1–5.
© 2006 *Springer.*

Prominent amongst the proprotein processing enzymes are the members of the family of subltilisin/kexin-like proprotein convertases (PCs). It took more than 15 years to identify these serine proteinases that can be subdivided into three sub-families: [A] The basic amino acid specific kexin-like PCs include seven members: PC1/3, PC2, Furin, PC4, PC5/6, PACE4 and PC7 [2]; [B] The pyrolysin-like subtilisin-kexin isoform SKI-1/S1P, also known as site 1 protease S1P [3]; and [C] The proteinase K-like neural apoptosis regulated convertase NARC-1/PCSK9 [4]. The last two convertases cleave at non-basic residues and process precursors that are distinct form those of the basic amino acid-specific convertases [3–6].

The discovery of these convertases from 1989–2003, elicited a wide interest in the scientific community as it was realized that these enzymes play key roles in various homeostatic as well as pathogenic events [2, 5–10]. The most evident role came from studies of the tumorigenic potential of these convertases, where it was shown that overexpression of one or more of the basic amino acid specific PCs leads to increased cell proliferation and enhanced metastasis, while their inhibition reverses this effect [11–14]. However, this is not universally the case, as a decreased expression of the Cys-rich domain containing PC5 [15, 16] and PACE4 [17] has been observed in various cancers including breast and ovarian cancers, as well as the increased metastatic potential of the human colon carcinoma HT-29 cells overexpressing α1-PDX, a potent inhibitor of the constitutively secreted conver-tases [18].

On another front, the implication of the PCs in viral infections became apparent from the processing sites of the surface glycoproteins of infectious viruses and of bacterial toxins [19]. In fact, data on various infectious viruses and bacterial toxins showed that cleavage of surface/spike glycoprotein precursors of these pathogens by one or more member of the PC-family, including the basic amino acid- specific Furin, PC7, PACE4 and/or PC5 (2) and the pyrolysin-like SKI-1/S1P (20) is a required step for the acquisition of fusiogenic potential and thus for their infectious and/or cell-cell spreading capacity [19, 21].

Recently, some of the convertases such as PC5/6, SKI-1/S1P and NARC-1/PCSK9, were implicated in cardiovascular complications. Examples include the vital role of SKI-1/S1P in the regulation of the synthesis of cholesterol and fatty acids via the cleavage within the Golgi of the two master switches of sterol, and fatty acid metabolism, the sterol regulatory element binding proteins [SREBP-1 and SREBP-2] [22, 23]. The convertase PC5/6 has also been implicated in vascular remodeling and the development of atherosclerosis [24, 25], as well as in the phenomenon known as restenosis that occurs following balloon angioplasty or stint implantation [26]. In addition, PC5/6, which is highly expressed in endothelial cells [27, 28] has been implicated in the activation of endothelial lipase, and hence could positively regulate the level of high density lipoproteins (HDL) [29].

Finally, the last member of the family NARC-1/PCSK9 has clearly been associated with the development of dyslipidemias, as specific mutations in its coding sequence are directly responsible for the development of a dominant form of either familial hyper-cholesterolemia [5] or hypo-cholesterolemia [30]. This is the first case of a dominant disease associated with mutations in one of the PCs. It seems that these mutations [6] result in either a gain/enhancement of an existing function, for those causing hyper-cholesterolemia [5], or in a loss of function in hypo-cholesterolemia patients [30]. The mechanism behind these pathologies is essentially related to one of the major roles of NARC-1/PCSK9 which is to enhance the degradation of the low density lipoprotein receptor (LDLR) [31] through a mechanism requiring entry into low pH endocytotic vesicles [32]. This exciting development opens the way to the development of anti-cholesterogenic drugs that could supplement the widely prescribed HMG-CoA reductase inhibitors, known as "statins" that themselves upregulate the expression of NARC-1/PCSK9 [33]. Indeed, supplementation of statins to the diet of mice lacking the expression of *PCSK9*, resulted in a marked additional decrease in the level of circulating total cholesterol [34].

The present monogram deals with multiple aspects of the proprotein convertases, from their discovery, to their analysis and to the projected pharmacological and clinical applications that may result from the inhibition of these enzymes. Thus, this is one example of "bench to bedside" directly applicable to the convertases. It is hoped that the use of modern day multiplexing technologies including various RNA and protein/peptide arrays should result in the development of specific convertase inhibitors that should find applications to control a wide variety of pathologies, including cancer and associated metastasis as well as dyslipidemias such as atherosclerosis and hypercholeste-rolemia. The importance of the PCs in the self renewal and maintenance of cancer stem cells [35] is a future area that begs extensive investigation, as it may opens the door towards stem cell-specific targeting of convertase inhibition. It took more than 30 years to unravel some of the mysteries of the proprotein convertases. It is hoped that the next decade will consolidate and expand the genetic, cellular and molecular knowledge of the PCs, including their 3D structures [36], in order to rationally design potent drugs that regulate their levels and/or activities *in vivo*.

REFERENCES

[1] Puente XS, Sanchez LM, Overall CM, Lopez-Otin C (2003) Human and mouse proteases: A comparative genomic approach. Nat Rev Genet **4**:544–558

[2] Seidah NG, Chretien M (1999) Proprotein and prohormone convertases: A family of subtilases generating diverse bioactive polypeptides. Brain Res **848**: 45–62

[3] Seidah NG, Prat A (2002) Precursor convertases in the secretory pathway, cytosol and extracellular milieu. Essays Biochem **38**:79–94

[4] Seidah NG, Benjannet S, Wickham L, Marcinkiewicz J, Jasmin SB, Stifani S, Basak A, Prat A, Chretien M (2003) The secretory proprotein convertase neural apoptosis-regulated convertase 1 (NARC-1): Liver regeneration and neuronal differentiation. Proc Natl Acad Sci USA **100**:928–933

[5] Abifadel M, Varret M, Rabes JP, Allard D, Ouguerram K, Devillers M, Cruaud C, Benjannet S, Wickham L, Erlich D, Derre A, Villeger L, Farnier M, Beucler I, Bruckert E, Chambaz J, Chanu B, Lecerf JM, Luc G, Moulin P, Weissenbach J, Prat A, Krempf M, Junien C, Seidah NG, Boileau C (2003) Mutations in PCSK9 cause autosomal dominant hypercholesterolemia. Nat Genet **34**:154–156

[6] Attie AD, Seidah NG (2005) Dual regulation of the LDL receptor—some clarity and new questions. Cell Metab **1**:290–292

[7]   Seidah NG, Mowla SJ, Hamelin J, Mamarbachi AM, Benjannet S, Toure BB, Basak A, Munzer JS,
      Marcinkiewicz J, Zhong M, Barale JC, Lazure C, Murphy RA, Chretien M, Marcinkiewicz M
      (1999) Mammalian subtilisin/kexin isozyme SKI-1: A widely expressed proprotein convertase
      with a unique cleavage specificity and cellular localization. Proc Natl Acad Sci USA
      96:1321–1326

[8]   Fugere M, Day R (2005) Cutting back on pro-protein convertases: The latest approaches to
      pharmacological inhibition. Trends Pharmacol Sci **26**:294–301

[9]   Khatib AM, Siegfried G, Chretien M, Metrakos P, Seidah NG (2002) Proprotein convertases in
      tumor progression and malignancy: Novel targets in cancer therapy. Am J Pathol **160**:1921–1935

[10] López de Cicco R, Bassi DE, Zucker S, Seidah NG, Klein-Szanto AJ (2005) Human carcinoma
      cell growth and invasiveness is impaired by the propeptide of the ubiquitous proprotein convertase
      furin. Cancer Res **65**:4162–4171

[11] Bassi DE, Mahloogi H, Klein-Szanto AJ (2000) The proprotein convertases furin and PACE4 play
      a significant role in tumor progression. Mol Carcinog **28**:63–69

[12] Cheng M, Xu N, Iwasiow B, Seidah N, Chretien M, Shiu RP (2001) Elevated expression of
      proprotein convertases alters breast cancer cell growth in response to estrogen and tamoxifen.
      J Mol Endocrinol **26**:95–105

[13] Khatib AM, Siegfried G, Prat A, Luis J, Chretien M, Metrakos P, Seidah NG (2001) Inhibition
      of proprotein convertases is associated with loss of growth and tumorigenicity of HT-29 human
      colon carcinoma cells: Importance of insulin-like growth factor-1 (IGF-1) receptor processing in
      IGF-1-mediated functions. J Biol Chem **276**:30686–30693

[14] Siegfried G, Basak A, Cromlish JA, Benjannet S, Marcinkiewicz J, Chretien M, Seidah NG,
      Khatib AM (2003) The secretory proprotein convertases furin, PC5, and PC7 activate VEGF-C to
      induce tumorigenesis. J Clin Invest **111**:1723–1732

[15] Nour N, Mayer G, Mort JS, Salvas A, Mbikay M, Morrison CJ, Overall CM, Seidah NG (2005)
      The Cysteine-rich domain of the secreted proprotein convertases PC5A and PACE4 functions
      as a cell surface anchor and interacts with tissue inhibitors of metalloproteinases. Mol Biol Cell
      16:5215–5226

[16] Cheng M, Watson PH, Paterson JA, Seidah N, Chretien M, Shiu RP (1997) Pro-protein convertase
      gene expression in human breast cancer. Int J Cancer **71**:966 971

[17] Fu Y, Campbell EJ, Shepherd TG, Nachtigal MW (2003) Epigenetic regulation of proprotein
      convertase PACE4 gene expression in human ovarian cancer cells. Mol Cancer Res **1**:569–576

[18] Nejjari M, Berthet V, Rigot V, Laforest S, Jacquier MF, Seidah NG, Remy L, Bruyneel E, Scoazec
      JY, Marvaldi J, Luis J (2004) Inhibition of proprotein convertases enhances cell migration and
      metastases development of human colon carcinoma HT-29 cells in a rat model. Am J Pathol
      164:1925–1933

[19] Thomas G (2002) Furin at the cutting edge: From protein traffic to embryogenesis and disease.
      Nat Rev Mol Cell Biol **3**:753–766

[20] Lenz O, ter Meulen J, Klenk HD, Seidah NG, Garten W (2001) The Lassa virus glycoprotein
      precursor GP-C is proteolytically processed by subtilase SKI-1/S1P. Proc Natl Acad Sci USA
      98:12701–12705

[21] Bergeron E, Vincent MJ, Wickham L, Hamelin J, Basak A, Nichol ST, Chretien M, Seidah
      NG (2005) Implication of proprotein convertases in the processing and spread of severe acute
      respiratory syndrome coronavirus. Biochem Biophys Res Commun **326**:554–563

[22] Brown MS, Ye J, Rawson RB, Goldstein JL (2000) Regulated intramembrane proteolysis: A control
      mechanism conserved from bacteria to humans. Cell **100**:391–398

[23] Pullikotil P, Vincent M, Nichol ST, Seidah NG (2004) Development of protein-based inhibitors of
      the proprotein of convertase SKI-1/S1P: Processing of SREBP-2, ATF6, and a viral glycoprotein.
      J Biol Chem **279**:17338–17347

[24] Stawowy P, Kallisch H, Kilimnik A, Margeta C, Seidah G, Chretien M, Fleck E, Graf K (2004)
      Proprotein convertases regulate insulin-like growth factor 1-induced membrane-type 1 matrix
      metalloproteinase in VSMCs via endoproteolytic activation of the insulin-like growth factor-1
      receptor. Biochem Biophys Res Commun **321**:531–538

[25] Stawowy P, Marcinkiewicz J, Graf K, Seidah N, Chretien M, Fleck E, Marcinkiewicz M (2001) Selective expression of the proprotein convertases furin, pc5, and pc7 in proliferating vascular smooth muscle cells of the rat aorta in vitro. J Histochem Cytochem **49**:323–332

[26] Veinot JP, Prichett-Pejic W, Picard P, Parks W, Schwartz R, Seidah NG, Chretien M (2004) Implications of proprotein Convertase 5 (PC5) in the arterial restenotic process in a porcine model. Cardiovasc Pathol **13**:241–250

[27] Beaubien G, Schafer MK, Weihe E, Dong W, Chretien M, Seidah NG, Day R (1995) The distinct gene expression of the pro-hormone convertases in the rat heart suggests potential substrates. Cell Tissue Res **279**:539–549

[28] Essalmani R, Marcinkiewicz E, Chamberland A, Mbikay M, Chretien M, Seidah NG, Prat A (2006) Genetic deletion of PC5/6 leads to early embryonic lethality. Mol Cell Biol **26**:354–61

[29] Jin W, Fuki IV, Seidah NG, Benjannet S, Glick JM, Rader DJ (2005) Proprotein covertases are responsible for proteolysis and inactivation of endothelial lipase. J Biol Chem **280**:36551–36559

[30] Cohen J, Pertsemlidis A, Kotowski IK, Graham R, Garcia CK, Hobbs HH (2005) Low LDL cholesterol in individuals of African descent resulting from frequent nonsense mutations in PCSK9. Nat Genet **37**:161–165

[31] Maxwell KN, Breslow JL (2004) Adenoviral-mediated expression of Pcsk9 in mice results in a low-density lipoprotein receptor knockout phenotype. Proc Natl Acad Sci USA **101**:7100–7105

[32] Benjannet S, Rhainds D, Essalmani R, Mayne J, Wickham L, Jin W, Asselin MC, Hamelin J, Varret M, Allard D, Trillard M, Abifadel M, Tebon A, Attie AD, Rader DJ, Boileau C, Brissette L, Chretien M, Prat A, Seidah NG (2004) NARC-1/PCSK9 and Its Natural Mutants: Zymogen cleavage and effects on the low density lipoprotein (LDL) receptor and LDL cholesterol. J Biol Chem **279**:48865–48875

[33] Dubuc G, Chamberland A, Wassef H, Davignon J, Seidah NG, Bernier L, Prat A (2004) Statins upregulate PCSK9, the gene encoding the proprotein convertase neural apoptosis-regulated convertase-1 implicated in familial hypercholesterolemia. Arterioscler Thromb Vasc Biol **24**:1454–1459

[34] Rashid S, Curtis DE, Garuti R, Anderson NN, Bashmakov Y, Ho YK, Hammer RE, Moon YA, Horton JD (2005) Decreased plasma cholesterol and hypersensitivity to statins in mice lacking Pcsk9. Proc Natl Acad Sci USA **102**:5374–5379

[35] Wang JC, Dick JE (2005) Cancer stem cells: Lessons from leukemia. Trends Cell Biol **15**:494–501

[36] Henrich S, Cameron A, Bourenkov GP, Kiefersauer R, Huber R, Lindberg I, Bode W, Than ME (2003) The crystal structure of the proprotein processing proteinase furin explains its stringent specificity. Nat Struct Biol **10**:520–526

CHAPTER 1

DISCOVERY OF THE PROPROTEIN CONVERTASES AND THEIR INHIBITORS

ABDEL-MAJID KHATIB[1], NATHALIE SCAMUFFA[1], FABIEN CALVO[1], MICHEL CHRÉTIEN[2] AND NABIL G. SEIDAH[3]

[1] INSERM U 716/AVENIR, Institut de Génétique Moléculaire, Paris, France
[2] Regional Protein Chemistry Centre, Diseases of Ageing Unit, Ottawa Health Research Institute, Loeb Building, 725 Parkdale Ave., Ottawa, Ontario, Canada
[3] Laboratory of Biochemical Neuroendocrinology, Institut de Recherches Cliniques de Montréal, 110 pine Ave West, Montreal, Quebec, Canada

Abstract: The members of the convertase family play a central role in the processing of various protein precursors ranging from hormones and growth factors to viral envelope proteins and bacterial toxins. The proteolysis of these precursors that occurs at basic residues is mediated by the proprotein convertases (PCs), namely: PC1, PC2, Furin, PACE4, PC4, PC5 and PC7. The proteolysis at non-basic residues is performed by subtilisin/kexin-like isozyme-1 (S1P/SKI-1) and the newly identified neural apoptosis-regulated convertase-1 (NARC-1/PCSK9). These proteases have key roles in many physiological processes and various pathologies including cancer, obesity, diabetes, neurodegenerative diseases and autosomal dominant hypercholesterolermia. Here we summarize the discovery of the proprotein convertases and their inhibitors, discuss their properties, roles, resemblance and differences

Keywords: Proprotein convertases, SKI-1/S1P, NARC-1/PCSK9, Prosegments, α1-PDX, 7B2, ProSAAS

1. PROPROTEIN CONVERTASES (PCs)

To date, seven mammalian members of subtilisin-related PCs that process substrates at basic residues have been identified. These include Furin/*PACE*, PC1/*PC3*, PC2, PC4, PACE4, PC5/*PC6*, and PC7/*LPC/PC8/SPC7* (Figure 1).

This somewhat confusing nomenclature arose from the simultaneous discovery of some of these enzymes by different groups. PCs are multi-domain serine proteinases

A-Majid Khatib (ed.), Regulation of Carcinogenesis, Angiogenesis and Metastasis by the Proprotein Convertases, 7–26.
© 2006 *Springer.*

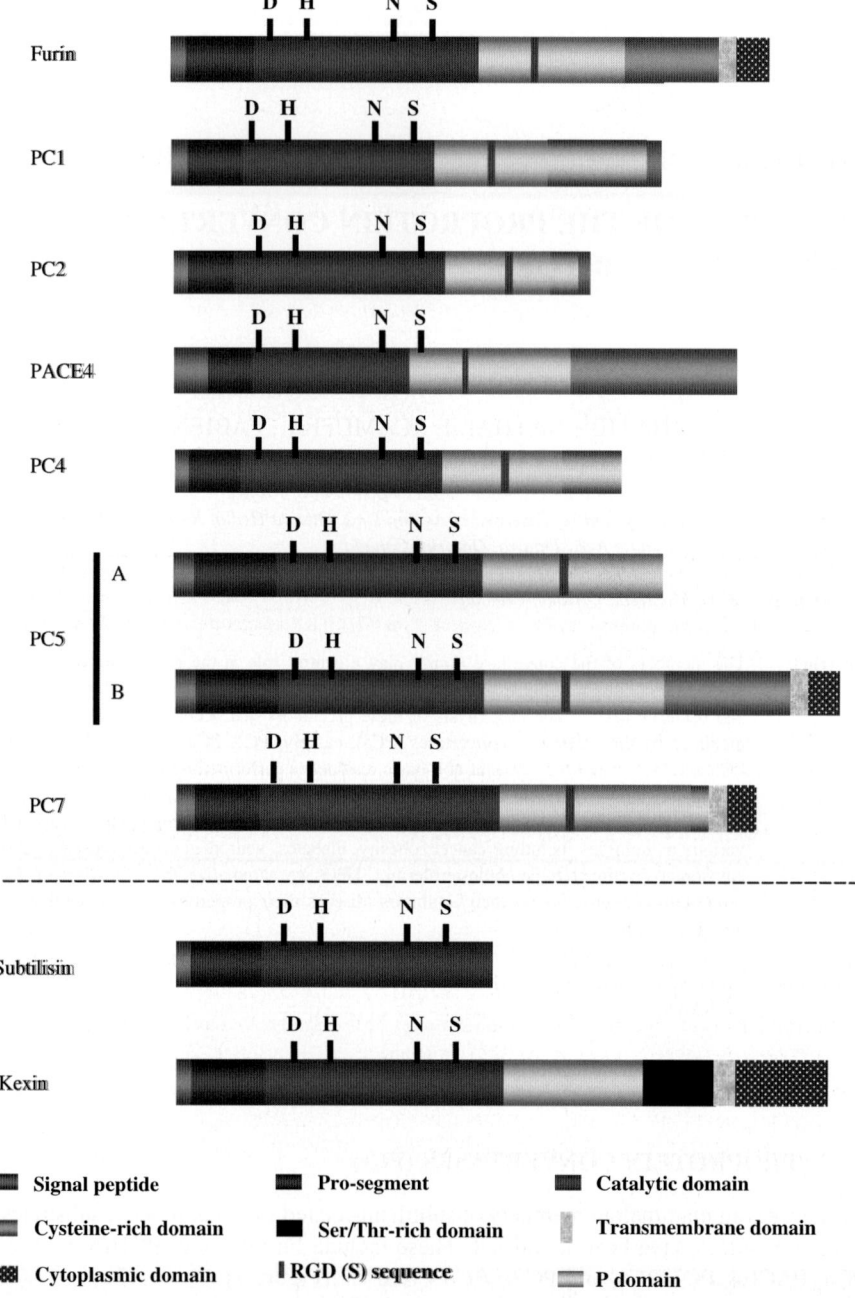

Figure 1. Schematic representation of the prohormone convertases PC1, PC2, Furin, PACE4, PC4, PC5 (A and B isoforms) and PC7. These PCs are multi-domain serine proteinases consisting of a signal peptide followed by prosegment, catalytic, middle, and cytoplasmic domains. Homology is highest in the catalytic domains and lowest in the carboxyl-terminal domains. The schematic representation for Kexin and subtilisin are given for comparison

Convertases	Amino acid number	Autocatalytic site	Accession number
Furin	794	[101]A-K-R-R-T-K-R-D	NP_002560
PC1	751	[105]K-E-R-S-K-R-S-V	P21662
PC2	638	[103]G-F-D-R-K-K-R-G	P16519
PACE4	969	[141]Q-E-V-K-R-R-V-K	P29122
PC4	654	[105]R-R-R-V-K-R-S-L	A54306
PC5	1870	[109]V-K-K-R-T-K-R-D	Q04592
PC7	785	[134]R-L-L-R-R-A-K-R	NP_004707
SKI-1	1052	[131]K-V-F-R-S-L-K-Y	NP_003782
NARC-1	692	[145]E-D-S-S-V-F-A-Q	NM_174936

Figure 2. Amino acid sequences of the autocatalytic sites of the PCs. Like their substrates, the pro-segments of the PCs are removed at sites cleaved by the PCs. Indicated are the number of amino acid and accession number for every PC

consisting of a signal peptide followed by prosegment, catalytic, middle, and cytoplasmic domains (Figure 1). Homology is highest in the catalytic domains and lowest in the carboxyl-terminal domains.

These enzymes cleave precursor proteins at basic residues within the general motif (K/R)-$(X)_n$-$(K/R)\downarrow$, where n = 0, 2, 4 or 6 and X any amino acid except Cys [1–4]. Usually most of the PCs cleave their substrates at pairs of basic amino acids, but several of them, with monobasic sites are also cleaved [1–4]. Some PCs, such as PC1, PC2 and PC5A, are sorted and activated in the regulated secretory pathway and thus process protein precursors whose secretion is regulated. In contrast, the trans-membrane proteins Furin, PACE4, PC5B and PC7 (Figure 1), cycle between the cell surface and the *trans* Golgi Network (TGN) and are involved in the processing of precursor proteins in the constitutive secretory pathway [1–4]. Like their substrates, the pro-segments of the PCs are also removed at a cleavage site containing a basic–amino acid PC motif (Figure 2), befitting their autoactivation [1–4].

1.1   Furin

Furin was the first convertase identified. Its discovery was made just after the availability of the Kex2 cDNA sequence. Kex2 is a cellular processing endopro-tease that is required for cleavage at dibasic sites within the killer toxin and the mating pheromone, α–factor precursors [5, 6]. In 1989, in an effort to find other

Convertases/ Inhibitors	Human		Mouse	
	Chromosomes	Cytogenetic	Chromosomes	Cytogenetic
Furin	15	15q26.1	7	7 D1-E2
PC1	5	5q15-q21	13	13C2
PC2	20	20p11.2	2	2G1
PACE4	15	15q26	7	7B5
PC4	19	19p13.3	10	10C1
PC5	9	9q21.3	19	19B
PC7	11	11q23.3	9	9A5.2
SKI-1	16	16q24	8	8E1
NARC-1	1	1p32.3	4	4C7
7B2	15	15q13-q14	2	2E5
Prosaas	X	Xp11.23	X	XA1.1

Figure 3. Chromosome localisation of mouse and human PCs and the inhibitors Prosaas and 7B2. Note the approximate position of PACE4 gene to fur gene on the human chromosome 15 and mouse chromosome 7 suggesting their probable common ancestry by gene duplication

related Kex2 enzymes, Fuller et al., identified the first mammalian homologue of Kex2 [7], *fur* gene. This gene is located on the human chromosome 15 and on mouse chromosome 7 (Figure 3). The *Furin* gene (*PCSK3*) was unexpectedly discovered by Roebroek et al., a few years earlier due to its proximity to the c-fes/fps protooncogene (*fur* being: *c-fes/fps* **u**pstream **r**egion) [8]. At that time the product of the *fur* gene was believed to be a growth factor receptor because of the presence of a cysteine-rich domain and a putative trans-membrane domain in its sequence (Figure 1, [9]). Subsequently, the cloning of full-length Furin cDNA revealed that Furin was structurally analogous to Kex2, although the Ser/Thr-rich domain in Kex2 was replaced by a cysteine-rich domain (Figure 1, [10]). Furin is a membrane protein, initially produced as a 104 kDa pro-furin precursor which is rapidly converted into a 98 kDa form by an autocatalytic process (Figure 2, [11, 12].) This autocatalytic cleavage of the pro-Furin occurs in the endoplasmic reticulum (ER) and is a perquisite for the exit of the mature Furin molecule out of the ER to reach the cell surface [13, 14]. Unlike most other convertases, Furin has a widespread distribution being present in all tissues and cells examined so far.

1.2   PC1 and PC2

In an effort to find additional Furin-like enzymes, the polymerase chain reaction was used successfully to detect and amplify conserved sequences within the catalytic domain of Furin and Kex2. In 1990, Seidah et al., identified, in mouse pituitary, the cDNA of two additional PC-related enzymes that were called PC1 and PC2 [15]. At approximately the same time, Smeekens and Steiner identified in human insulinoma a cDNA coding for PC2 [16]. The human and mouse *PC1* genes (*PCSK1*) are localized on chromosomes 5 and 13, respectively, whereas the *PC2* gene (*PCSK2*) is localized on human chromosome 20 and on mouse chromosome 2 (Figure 3). The corresponding protein of the full-length cDNA of PC1 is a 751-residue protein and the cDNA of PC2 encodes a 638-residue protein. Contrary to Kex2 and Furin, both PC1 and PC2 lack a transmembrane domain (Figure 1) [15, 16]. In 1991, using similar approaches, Smeekens et al., identified a PC-related enzyme highly expressed in the mouse AtT20 anterior pituitary cell line that unfortunately was called PC3 [17], since it turned out to be identical to PC1 [18, 19, 20]. Studies in various laboratories revealed that PC1 and PC2 process peptide hormones and neuropeptide precursors within the dense core vesicles of the regulated secretory pathway of the brain and the neuroendocrine system [21, 22]. Although PC1 and PC2 are structurally very similar, each convertase has definite substrate site preferences. Among the major substrates of these enzymes are proopiomelanocortin (POMC), proinsulin and proglucagon [23, 24]. Regulation of the activity of PC1 occurs by both its N-and C-terminal domains. Following its N-terminal autocatalytic cleavage within the endoplasmic reticulum, the 84 kDa form of PC1 is transported to the *trans* Golgi Network (TGN) and secretory granules to undergo two other autocatalytic cleavages, one within the inhibitory prosegment and the other at its carboxy-terminal domain to generate the fully active 66-kDa form [25], the major form found in islets of Langerhans and in secretory granules of AtT20 cells [25]. Although PC2 is also autocatalytically processed prior to activation like PC1 (Figure 2), the removal of its prosegment is less efficient and PC2 slowly exits from the ER as a zymogen (proPC2) and is processed to PC2 only in immature secretory granules. This difference in the time course of activation of PC1 and PC2 was reportedly linked to pH and calcium levels [26]. The cleavage of proPC1 to the 84 kDa PC1 occurs at a neutral pH and is calcium-independent, whereas PC2 is activated much more slowly in the immature secretory granules at pHs 5–6 in a calcium-dependent fashion [26]. As a consequence of the different temporal activation of PC1 and PC2 in cells expressing both enzymes, PC1 will cleave precursors before PC2, leading to an ordered cleavage mechanism that may explain the first cleavage of POMC into β-LPH and then into β-endorphin, peptide products that require the consecutive action of PC1 and PC2, respectively [27, 28].

1.3   PACE4

With a polymerase chain reaction methodology similar to the one used for the identification of Furin, PC1 and PC2, Kiefer et al., identified the convertase PACE4 using specific primers for the paired basic amino acid residue processing motifs of the available PCs [29]. PACE4 contains distinct features that are not present in the previously identified three convertases. These include an extended signal peptide region and large carboxyl-terminal cysteine-rich region (Figure 1) [29, 30]. PACE4 is expressed in most tissues, with highest levels occurring in the liver [29]. It processes a variety of substrates [30]. Like other PCs, the maturation of proPACE4 occurs via an intramolecular autocatalytic cleavage of its propeptide (Figure 2). This is the rate-limiting step for the secretion of the mature PACE4 [31, 32]. Furthermore, the secretion and the maturation of PACE4 are also controlled by the carboxy terminal sequence of PACE4 [31, 32]. Deletion of the last 25 residues of PACE4 has been shown to induce a marked acceleration in both the maturation and secretion of mature PACE4 [31]. Another property of PACE4 is its ability to bind heparan sulfate proteoglycans in the extracellular matrix (ECM) [33]. The PACE4 heparin-binding region was localized in the cationic region of amino acids between residues 743 and 760. This suggests a spatial role for PACE4 in the regulation of the biological activities of its substrates [33]. Very recently, we have shown that the C-terminal Cys-rich domain of PACE4 anchors the secreted enzyme to the plasma membrane *via* a complex with one or more member of the tissue inhibitor of metalloproteases (TIMPs) through binding of the complex to cell surface heparan sulfate proteoglycans [34]. Localization of the *PACE4* gene (*PCSK6*) revealed its closeness to the *fur* gene on the human chromosome 15 and mouse chromosome 7 (Figure 3), suggesting a probable common ancestry by gene duplication [29].

Despite a likely common origin, the regulation of Furin and PACE4 expression appears quite different. While both are up-regulated by phorbol 12-myristate 13-acetate (PMA) and tumor necrosis factor (TGF), PACE4 is also upregulated by platelet derived growth factor-BB (PDGF-BB), indicating a unique role for PACE4 in platelet production [35, 36]. Recent studies revealed that the expression of PACE4 is down-regulated by the basic helix-loop-helix transcription factors hASH-1 and MASH-1, suggesting co-regulation of PACE4 and its substrates by these transcription factors [37].

1.4   PC4

Like other PCs, identification of PC4 was based on PCR strategies and was simultaneously identified from mouse testis by Nakayama and our group [38, 39]. It is a 654-residue protein, which possesses the same subtilisin-like catalytic domain found in Furin, PC1, PC2, and Kex2 (Figure 1). Distribution analysis in various cell lines and tissues revealed that PC4 appears to be exclusive to testis and ovarian cells [38–41]. Northern blot analysis indicates that PC4 mRNA is detectable only in the testis after the 20th day of postnatal development and was primarily expressed in round spermatids, suggesting that PC4 is involved in the maturation of precursor proteins found in testicular germ cells. Subsequently, the importance of PC4 in these processes was shown by PC4 gene expression during spermatogenesis [38–40]. Although PC4 is able to efficiently process various protein precursors in the testis, a specific substrate for PC4 expressed only in this organ remains unknown.

The *PC4* gene (*PCSK4*) is located on chromosome 19 and 10 in human and mouse, respectively (Figure 3).

1.5   PC5 (Isoforms PC5A and PC5B)

The 915 amino acid isoform PC5A was identified and cloned by our group using RT-PCR and oligonucleotide sequences derived from conserved sequences of PC1, PC2, Furin, and PC4, in both mouse and rat tissues [41]. The same year, the group of Nakagawa et al., cloned this convertase and named it PC6 [42]. The *PC5* gene (*PCSK5*) is localized on human chromosome 9 and mouse chromosome 19 (Figure 3). The human *PCSK5* gene encodes two isoforms: the 915 amino acid PC5A and a C-terminally extended 1870-residue protein (PC5B) with multiple Cys-rich domains. Both isoforms contain a subtilisin-like catalytic domain and PC5A exhibits a high similarity to PACE4, especially at the COOH-terminal Cys-rich region (Figure 1) [42]. Northern blot analysis revealed that PC5 mRNA, as with Furin and PACE4 mRNA, was expressed in various tissues and cell lines [42–45]. Its highest expression is in adrenal cortex and small intestine suggesting possible roles in stress response and in processing protein substrates of gastrointestinal peptides [42–45]. Like PACE4, the expression of PC5 is upregulated by PDGF-BB and during cell proliferation [44]. Many substrates have been reported to be efficiently processed by PC5; including growth factors such PDGF-A [45], PDGF-B [46] and VEGF-C [47], receptors such as IGFI-1 receptor (1) and various integrins [48]. While these substrates were also shown to be processed by other PCs, certain precursor proteins are processed effectively mostly by PC5, such as neural adhesion molecule L1 [49] and Lefty protein [50]. Similar to other PCs, the activity and secretion of PC5 is also regulated by its prosegment. The pro-region of PC5 was shown to prevent IGF-1 receptor (1) and VEGF-C processing by PC5, both *in vitro* and *in vivo* [47, 51] suggesting an inhibitory role of the PC5 propeptide.

1.6   PC7

This convertase was identified in 1996 by our group [52], Bruzzaniti et al. [53] and Meerabux et al. [54]. Meerabux identified PC7 through its involvement in a chromosome translocation that occurred in a particular lymphoma [54]. This translocation is the result of a fusion between an intron in the 3′-untranslated region of PC7 with a sequence close to the switch region S gamma 4 of the IGH locus. The product of the *PC7* gene (*PCSK7*) encodes a 785 residue protein with a large homology to all members of the PC family (Figure 1). Using PCR and degenerate primers to conserved amino acid residues in the catalytic region of the PCs, Bruzzanti et al., predicted the product of the gene they identified (called PC8) to be 785 residues [53]. The catalytic region of this protein is more than 50% identical in primary sequence to the other PCs. Using similar technologies, we also isolated a cDNA coding for a gene from the rat anterior pituitary that we named PC7. We found the open reading frame codes for a prepro-PC with a 36-amino acid signal peptide, a 104-amino acid prosegment, and a 747-amino acid type I membrane-bound glycoprotein, representing the mature form of PC7 [52]. Distinct from Furin (*PCSK3*) and PACE4 (*PCSK6*) genes, both mapping to chromosome 15, PCSK7 maps to chromosome 11 (Figure 3). Phylogenetic analysis suggested that PC7 is the most ancestral member of the seven basic amino acid-specific proprotein convertases [52]. Northern blot analyses demonstrated significant expression of PC7 mRNA in the colon and lymphoid-associated tissues. *In situ* hybridization and histochemistry analysis in various tissues revealed that PC7 co-localizes with Furin, suggesting widespread proteolytic functions of PC7 and its participation with Furin in the activation of several substrates [52–57].

2.   PROPROTEIN CONVERTASES THAT PROCESS SUBSTRATES AT NON-BASIC RESIDUES

2.1   Subtilisin/Kexin-like Isozyme-1 (SK-1)

In 1999, using reverse transcriptase-PCR and degenerate oligonucleotides, derived from the active-site residues of subtilisin/kexin-like serine proteinases, we identified in human, rat, and mouse, a type I membrane-bound proteinase, which we called subtilisin/kexin-isozyme-1 (SKI-1) [58]. It was so named because of the homology of its catalytic domain to the bacterial subtilisin BPN (Figure 4). In contrast to the basic amino acid-specific PCs, this convertase appears to prefer processing precursors at residues within the general motif $RX(V, I, L)(K, F, L)\downarrow$, with the preferred critical basic Arg/Lys and aliphatic (Leu/Ile/Val) residues occupying positions P4 and P2, respectively [58].

Data bank searches revealed that Sakai et al., also identified a few month earlier a similar hamster enzyme from CHO cells, which they named Site-1 protease (S1P). They determined that this enzyme was involved in the control of lipid metabolism by mediating the cleavage of Sterol Regulatory Element-Binding Proteins (SREBPs) in its luminal loop [59]. Previously, SREBPs were described to play a key role in the fundamental feedback mechanism of cellular lipid homeostasis.

The transcriptional activation of genes containing sterol responsive elements (SRE) is known to be regulated by sterols through modulation of the proteolytic maturation of SREBPs [59]. The two known SREBPs (SREBP1 and SREBP2) are inserted into the membrane of the endoplasmic reticulum envelope in a wide variety of tissues. In sterol-deficient cells, proteolytic cleavage of SREBPs by SKI-1 and S2-P protease releases their N-terminal mature form from the membrane into the cytosol enabling them to enter the nucleus (Figure 5), where they bind to the SREs and activate genes involved in the biosynthesis of cholesterol, triglycerides, and fatty acids [59]. In the presence of sterols, the proteolytic process is inhibited and the transcription of the genes is reduced [59] (Figure 5).

The gene of SKI-1/S1P (*PCSK8*) is located on human chromosome 16 and mouse chromosome 8 (Figure 3), and is expressed in most tissues and cells. To date, several viral glycoproteins in addition to SREBPs, as well as the brain-derived

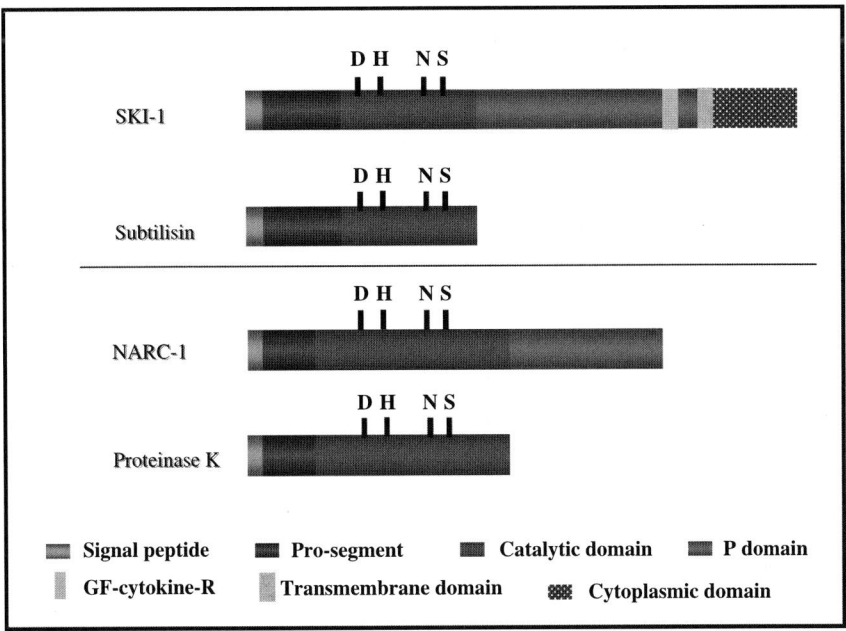

Figure 4. Schematic representation of the prohormone convertases SK-1 and NARC-1. The convertase subtilisin/kexin-isozyme-1 (SKI-1) possesses a catalytic domain with high homology to bacterial subtilisin BPN, whereas the neural apoptosis-regulated convertase-1 (NARC-1) belongs to the proteinase K-like subtilases

neurotrophic factor, ATF-6 and endocrine polypeptide somatostatin were found to be SKI-1 substrates [59–64]. New substrates include CREB-containing precursors, such as CREB-4 were also reported to be cleaved by SKI-1/S1P [65]. As with the PCs, the precursor protein of SKI-1 is also autocatalyticaly cleaved (Figure 2) and can be further processed into two membrane-bound forms of SKI-1 (120 and 106 kDa), differing by the nature of their N-glycosylation. Some of these SKI-1 forms are shed into the medium as a 98-kDa form.

2.2   Neural Apoptosis-regulated Convertase 1 (NARC-1/PCSK9)

Through a search of patent databases, using as a bait a small sequence of the conserved catalytic domain of SKI-1/S1P, we identified a protein belonging to proteinase K-like subtilases (Figure 4) called neural apoptosis-regulated convertase 1 (NARC-1) or PCSK9. NARC-1/PCSK9 was previously identified by two pharmaceutical companies [66], based on the cloning of up-regulated cDNAs after the induction of apoptosis by serum deprivation in the primary cerebellar neurons and by means of global cloning of secretory proteins [66]. Like other convertases, NARC-1/PCSK9 is also synthesized as a zymogen that undergoes autocatalytic intramolecular processing in the ER (Figure 2). This cleavage occurs within the

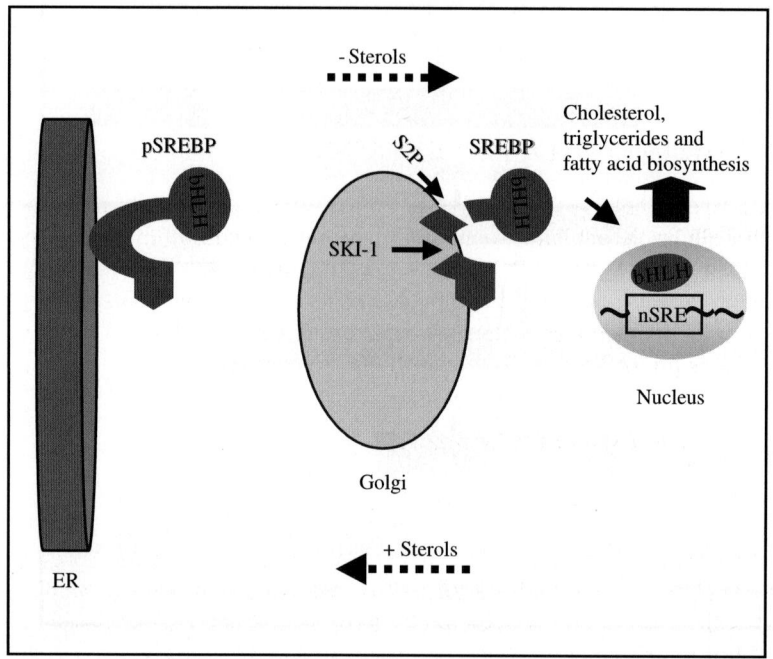

Figure 5. Role of SKI-1/S1P in the processing of SREBP. The sterol regulatory element binding protein precursors (SREBPs) are inserted into the membrane of the endoplasmic reticulum (ER) envelope in various tissues and the amino-terminal transcription-factor domain (bHLH-zip) is located in the cytoplasmic compartment. Under insufficient amount of sterols, the SREBP precursor protein travels to the Golgi apparatus where SKI-1/S1P cleaves at site-1 in the luminal loop and produce the substrate for the Site-2 protease (S2P), which cleaves at site-2. This second cleavage releases the transcription-factor domain from the membrane that enters the nucleus and induces the increased transcription of target genes. In the presence of sterols, the proteolytic process is inhibited and the transcription of the genes is reduced. bHLH-zip: basic helix-loop-helix leucine-zipper

motif **SSVFAQ SIP** [67]. Northern blots and *in situ* hybridization analyses revealed that in the adult NARC-1/PCSK9 mRNA expression is restricted to the liver, kidney and small intestine. Unlike PC7 and SKI-1, but similar to Furin, PC5 and PACE4, the mRNA of NARC-1/PCSK9 was up-regulated during liver regeneration following partial hepatectomy [68]. Overexpression of NARC-1/PCSK9 in primary culture of embryonic telencephalon cells at day 13.5 induced differentiation of neuronal progenitors, suggesting a role for NARC-1/PCSK9 in enhancing the differentiation/proliferation of cortical neurons [66]. Recently, we have shown that point mutations in human PCSK9 are associated with the development of severe hypercholesterolemia phenotypes [69], likely through a grain of function [70]. Conversely, other mutations resulting in early termination of the coding region (non-sense mutations) resulted in a loss of function and hence familial hypocholesterolemia [71]. Thus, mutations in PCSK9 results in a dominant form of either hypo-or hyper-cholesterolemia, suggesting that inhibitors of these enzymes may lead to novel pharmaceutical drugs to further lower circulating cholesterol levels as a supplement to the conventional HMG-CoA reductase inhibitors known as "statins".

3.   PROPROTEIN CONVERTASE INHIBITORS

To date, the propeptides or prosegments of the PCs constitute the only naturally occurring intracellular PC inhibitor found in the mammalian constitutive secretory pathway [1–4] and, in the case of PC1, its C-terminal domain [72]. Aside from the prosegment inhibitors, the activities of the regulated secretory pathway convertases PC1 and PC2 are also regulated by their selective and specific inhibitors/binding partners, known as proSAAS [73, 74] and 7B2 [75] respectively.

3.1   Naturally Occurring PC2 Inhibitor 7B2

In 1982, during the purification of the POMC N-terminal glyco-segment from pig anterior pituitaries, we discovered the protein 7B2 [75]. Subsequently, the homologues of this peptide were cloned in tissues and organs of other species, including human, and showed high homology between mammals [75–78]. Studies on the tissue distribution and secretion of 7B2 revealed its predominance in endocrine and neural tissues, including the brain and adrenal medulla, as well as the pituitary, thyroid and pancreas [75].

The gene for 7B2 is located on human chromosome 15 and mouse chromosome 2 (Figure 3). It is produced as an intracellular precursor of 25–29 kDa. This 7B2 precursor is converted into a secreted form of 18–21 kDa by PC cleavage after the $RRRRR^{155}$ motif, followed by carboxypeptidase E (CPE) removal of the 5 basic residues. After processing, 7B2 proteins are packaged into dense-core vesicles and are secreted upon exocytotic stimulation [75]. Pulse–chase studies showed that proPC2 is bound to pro7B2 in the early compartments of the secretory pathway dissociates from it in later ones and serves as an intracellular proPC2 chaperone that prevents the premature activation of the zymogen during its transit in the regulated secretory pathway [75]. Attachment of pro7B2 to proPC2 in the ER generates an inactive complex that is transported to the TGN where pro7B2 is cleaved into an N-terminal protein and an inhibitory C-terminal 31 aa peptide (CT-7B2). ProPC2 is then autocatalytically cleaved after the prodomain as the complex is transported into the immature secretory granules [75]. In the acidic environment of these organelles, the prodomain and 7B2 dissociate from the enzyme, which then cleaves the PC2-specific inhibitory CT-7B2 resulting in fully active PC2.

3.2   Naturally Occurring PC1 Inhibitor ProSAAS

ProSAAS was identified by Fricker et al. during an analysis of peptides not properly processed in Cpe^{fat}/Cpe^{fat} mice lacking carboxypeptidase E activity due to a point mutation in the carboxypeptidase E gene [79, 80]. These mice accumulate peptides with C-terminal Lys and/or Arg extensions. Using an affinity column, peptides with C-terminal basic residues from Cpe^{fat}/Cpe^{fat} tissues were isolated and analyzed. Five of these peptides were found to be encoded by proSAAS [81]. Subsequent overexpression of proSAAS in endocrine cells revealed its selective inhibitory effect on PC1 [81]. The proSAAS gene is located on the human and mouse chromosome X (Figure 3) and, similarly to 7B2, proSAAS is largely expressed in neuroendocrine cells and its inhibitory domains are located at the C terminus. In contrast to 7B2, which is required for the expression and secretion of active convertase PC2 [82–84], active PC1 can be expressed in cells lacking proSAAS [82–84]. Despite the absence of data on proSAAS null mice, taking together with its inhibitory role on PC1, and similarities to 7B2, proSAAS may be assumed to have other functions such as the control of the body mass blood glucose levels as recently revealed by analysis of transgenic mice expressing proSAAS [85].

3.3   Prosegments and Exogenous Inhibitors

Since the discovery of Furin, many attempts have been made to develop inhibitors to control the activity of the PCs. Initially, taking advantage of the fact that PCs are synthesized as inactive zymogens autocatalytically activated, Anderson et al., demonstrated that the prosegment of Furin, when used as a fusion protein to glutathione S-transferase, exhibits a potent *in vitro* inhibitory activity on Furin [86]. Previously, we found that purified prosegments and synthetic peptides derived from the prosegments of PC1, PC7 and Furin are potent inhibitors of their corresponding enzymes [87–91]. Using these inhibitors, we were able to intracellularly inhibit the processing of various PC substrates, including PDGF-A [45], PDGF-B [46] VEGF-C [47] and IGF-1 receptor (1.)

In addition to these naturally occurring inhibitors, many exogenous inhibitors were proposed to control the activity of the convertases. Of these molecules, the trypsin inhibitor and the third domain of turkey ovomucoid have been reported to be inhibitors for furin [92]. Subsequently, Garten et al. [93] have shown that acylated peptidyl chloromethane, containing a consensus furin cleavage sequence, decanoyl-Arg-Glu-Lys-Arg-COCH$_2$Cl, that inhibits Furin activity *in vitro* at low micromolar concentrations to block the cleavage of influenza-virus HA. While these inhibitors were useful for study of the processing of various proteins by Furin, they appear to be unstable and unable to completely block the processing of various PC substrates *in vivo* due to their inefficiencies and/or decreased capability in entering cells. In 1988, Bathurst et al., and Brennan et al., proposed the use of protein-based inhibitors to control the activity of PCs [94]. They demonstrated that the variant of α1-antitrypsin, called α1-anti-trypsin Pittsburgh (α1-PIT), which has a replacement of the reactive-site Met residue by Arg, inhibits, *in vitro*, the processing of proalbumin by Kex2p [94]. Subsequently, the group of G. Thomas developed another variant of α_1-antitrypsin, called α_1-anti-trypsin Portland (α_1-PDX), in which the reactive-site Ala-Ile-Pro-Met has been replaced by Arg-Ile-Pro-Arg. This serpin inhibits Furin in the subnanomolar range, three times lower than that α_1-PIT.

Kinetic analysis showed that a portion of bound α_1-PDX operates as a suicide inhibitor [94–97]. Once bound to Furin's active site, α_1-PDX can either undergo proteolysis by Furin or form a kinetically trapped SDS-stable complex with the enzyme. Furthermore, when expressed in cells, α_1-PDX, was shown to be a potent inhibitor of Furin-mediated cleavage of HIV gp160 [97], and subsequently demonstrated to inhibit all PCs involved in processing within the constitutive secretory pathway [1, 97–101]. Inhibition of PCs by α_1-PDX has been shown to reduce the production of the APPα [102] and block the activation of the pore-forming toxin proaerolysin [103], the maturation of infectious pathogens glycoproteins [97], the proteolytic activation of BMP4 [104] and the cleavage of IGF-1R [1, 105], PDGF-A [45], PDGF-B [46] and VEGF-C [47].

In an attempt to produce other PC inhibitors, researchers mutated the bait region of the general protease inhibitor α_2-macroglobulin (RVGFYESDVM^{690} into RVRSKRSLVM^{690}) [106]. This variant was reported to inhibit processing of several Furin substrates including HIV type 1 glycoprotein gp160, von Willebrand factor and TGF-β1 [106]. Other inhibitors were suggested, such as the ovalbumin-type serpin human proteinase inhibitor-8, which contains two instances of the minimal Furin recognition sequence (VVRNSRCSRM^{343}). Although this inhibitor was shown to inhibit Furin in a rapid and tight binding manner, it required the addition of a signal peptide before it could inhibit Furin *in vivo* [107]. Additionally, the hexa-D-arginine was reported to be a potent and relatively specific Furin inhibitor; however, it showed reduced ability to cross the cell membrane [108].

4.   SUMMARY AND CONCLUDING REMARKS

Since the discovery of Furin, the first mammalian convertase identified, cumulative knowledge has been acquired regarding the physiological and physiopathological role of these enzymes. The data obtained on the functional role of these enzymes by the use of null mice provided exceptional information, not only on the precursor proteins that are processed by one or more PCs, but also precious information on the importance of these enzymes in normal physiological situations. To date, based on the available PC-null mice, only the absence or dysfunction of Furin [109], PC5 [110] and SKI-1/S1P [111] are lethal at the embryonic stage. Mice with disrupted PC1or PC2 are viable despite their hormonal and/or neuro-endocrinal deficiency [112, 113]. PACE4 deficient animals show bone defects [114] and PC4 null mice are infertile or subfertile [115]. These varieties in the PC knockout phenotypes reveal the complexity and wide array of the protein precursors that are processed by these enzymes. Protein precursors may be processed by one specific convertase, a limited set or multiple convertases. The determination of the knockout phenotype observed in the PC-null mice seems to be more likely due to a defect in the processing of specific protein precursors by specific PCs.

While the PC null mice studies confirm the critical role of these enzymes in the activation of proteins involved in physiological processes, there is also growing evidence of their role in various pathological processes and diseases. Some PCs have been reported to be involved in Alzheimer's disease, rheumatoid arthritis, cancer and other pathologies. In this chapter, we have described the progress made in establishing potent and specific inhibitors to control PC activity. Some of these inhibitors, particularly α_1-PDX, were shown to dramatically reduce tumor growth and the malignant phenotype of various caner cells [1, 105]. α_1-PDX was also shown to inhibit the processing of the HIV-1 GP 160 protein and other viral glycoproteins and, in turn, the production of infectious viruses. Recently, inhibition of Furin by the inhibitor Dec-RVKR-CH(2)Cl was revealed to prevent cartilage degradation induced by cytokines, suggesting the inhibition of PCs as a potential therapeutic intervention in arthritic diseases [116].

ACKNOWLEDGEMENTS

This work was supported by the grant from the Fondation pour la Recherche Médicale and Avenir INSERM Award to AM K, Paris, France.

REFERENCES

[1] Khatib AM, Siegfried G, Chretien M, Metrakos P, Seidah NG (2002) Proprotein convertases in tumor progression and malignancy: Novel targets in cancer therapy. Am J Pathol **160**:1921–1935

[2] Nakayama K (1997) Furin: A mammalian subtilisin/Kex2p-like endoprotease involved in processing of a wide variety of precursor proteins. Biochem J **327**:625–635

[3] Zhou A, Webb G, Zhu X, Steiner DF (1999) Proteolytic processing in the secretory pathway. J Biol Chem **274**:20745–20748

[4] Seidah NG, Chretien M, Day R (1994) The family of subtilisin/kexin like pro-protein and pro-hormone convertases: Divergent or shared functions. Biochimie **76**:197–209

[5] Julius D, Brake A, Blair L, Kunisawa R, Thorner J (1984) Isolation of the putative structural gene for the lysine-arginine-cleaving endopeptidase required for processing of yeast prepro-alpha-factor. Cell **37**:1075–1089

[6] Mizuno K, Nakamura T, Ohshima T, Tanaka S, Matsuo H (1988) Yeast KEX2 genes encodes an endopeptidase homologous to subtilisin-like serine proteases. Biochem Biophys Res Commun **156**:246–254

[7] Fuller RS, Brake AJ, Thorner J (1989) Intracellular targeting and structural conservation of a prohormone-processing endoprotease. Science **246**:482–486

[8] Roebroek AJ, Schalken JA, Bussemakers MJ, van Heerikhuizen H, Onnekink C, Debruyne FM, Bloemers HP, Van de Ven WJ (1986) Characterization of human c-fes/fps reveals a new transcription unit (fur) in the immediately upstream region of the proto-oncogene. Mol Biol Rep **11**:117–125

[9] Roebroek AJ, Schalken JA, Leunissen JA, Onnekink C, Bloemers HP, Van de Ven WJ (1986) Evolutionary conserved close linkage of the c-fes/fps proto-oncogene and genetic sequences encoding a receptor-like protein. EMBO J **59**:2197–2202

[10] van de Ven WJ, Voorberg J, Fontijn R, Pannekoek H, van den Ouweland AM, van Duijnhoven HL, Roebroek AJ, Siezen RJ (1990) Furin is a subtilisin-like proprotein processing enzyme in higher eukaryotes. Mol Biol Rep **14**:265–275

[11] Leduc R, Molloy SS, Thorne BA, Thomas G (1992) Activation of human furin precursor processing endoprotease occurs by an intramolecular autoproteolytic cleavage. J Biol Chem **267**:14304–14308

[12] Creemers JW, Siezen RJ, Roebroek AJ, Ayoubi TA, Huylebroeck D, Van de Ven WJ (1993) Modulation of furin-mediated proprotein processing activity by site-directed mutagenesis. J Biol Chem **268**:21826–21834

[13] Takahashi S, Nakagawa T, Kasai K, Banno T, Duguay SJ, Van de Ven WJ, Murakami K, Nakayama K (1995) A second mutant allele of furin in the processing-incompetent cell line, LoVo. Evidence for involvement of the homo B domain in autocatalytic activation. J Biol Chem **270**:26565–26569

[14] Molloy SS, Thomas L, VanSlyke JK, Stenberg PE, Thomas G (1994) Intracellular trafficking and activation of the furin proprotein convertase: Localization to the TGN and recycling from the cell surface. EMBO J **13**:18–33

[15] Seidah NG, Gaspar L, Mion P, Marcinkiewicz M, Mbikay M, Chretien M (1990) cDNA sequence of two distinct pituitary proteins homologous to Kex2 and furin gene products: Tissue-specific mRNAs encoding candidates for pro-hormone processing proteinases. DNA Cell Biol **9**:415–424

[16] Smeekens SP, Steiner DF (1990) Identification of a human insulinoma cDNA encoding a novel mammalian protein structurally related to the yeast dibasic processing protease Kex2. J Biol Chem **265**:2997–3000

[17] Smeekens SP, Avruch AS, LaMendola J, Chan SJ, Steiner DF (1991) Identification of a cDNA encoding a second putative prohormone convertase related to PC2 in AtT20 cells and islets of Langerhans. Proc Natl Acad Sci U S A **88**:340–344

[18] Seidah NG, Marcinkiewicz M, Benjannet S, Gaspar L, Beaubien G, Mattei MG, Lazure C, Mbikay M, Chretien M (1991) Cloning and primary sequence of a mouse candidate prohormone convertase PC1 homologous to PC2, Furin, and Kex2: Distinct chromosomal localization and messenger RNA distribution in brain and pituitary compared to PC2. Mol Endocrinol **5**:111–122

[19] Korner J, Chun J, Harter D, Axel R (1991) Isolation and functional expression of a mammalian prohormone processing enzyme, murine prohormone convertase 1. Proc Natl Acad Sci U S A **88**:6834–6838

[20] Steiner DF, Smeekens SP, Ohagi S, Chan SJ (1992) The new enzymology of precursor processing endoproteases. J Biol Chem **267**:23435–23438

[21] Seidah NG, Benjannet S, Hamelin J, Mamarbachi AM, Basak A, Marcinkiewicz J, Mbikay M, Chretien M, Marcinkiewicz M (1999) The subtilisin/kexin family of precursor convertases. Emphasis on PC1, PC2/7B2, POMC and the novel enzyme SKI-1. Ann N Y Acad Sci **885**:57–74

[22] Bell ME, Myers TR, Myers DA (1998) Expression of proopiomelanocortin and prohormone convertase-1 and -2 in the late gestation fetal sheep pituitary. Endocrinology **139**:5135–5143

[23] Steiner DF, Rouille Y, Gong Q, Martin S, Carroll R, Chan SJ (1996) The role of prohormone convertases in insulin biosynthesis: Evidence for inherited defects in their action in man and experimental animals. Diabetes Metab **22**:94–104

[24] Mineo I, Matsumura T, Shingu R, Namba M, Kuwajima M, Matsuzawa Y (1995) The role of prohormone convertases PC1 (PC3) and PC2 in the cell-specific processing of proglucagon. Biochem Biophys Res Commun **207**:646–651

[25] Vindrola O, Lindberg I (1992) Biosynthesis of the prohormone convertase mPC1 in AtT-20 cells. Mol Endocrinol **6**:1088–1094

[26] Shennan KI, Taylor NA, Jermany JL, Matthews G, Docherty K (1995) Differences in pH optima and calcium requirements for maturation of the prohormone convertases PC2 and PC3 indicates different intracellular locations for these events. J Biol Chem **270**:1402–1407

[27] Crine P, Gossard F, Seidah N, Blanchette L, Lis M, Chretien M (1979) Concomitant synthesis of beta-endorphin and alpha-melanotropin from two forms of pro-opiomelanocortin in the rat pars intermedia. Proc Natl Acad Sci U S A **76**:5085–5089

[28] Benjannet S, Rondeau N, Day R, Chretien M, Seidah NG (1991) PC1 and PC2 are proprotein convertases capable of cleaving proopiomelanocortin at distinct pairs of basic residues. Proc Natl Acad Sci U S A **88**:3564–3568

[29] Kiefer MC, Tucker JE, Joh R, Landsberg KE, Saltman D, Barr PJ (1991) Identification of a second human subtilisin-like protease gene in the fes/fps region of chromosome 15. DNA Cell Biol **10**:757–769

[30] Laprise MH, Grondin F, Cayer P, McDonald PP, Dubois CM (2002) Furin gene (fur) regulation in differentiating human megakaryoblastic Dami cells: Involvement of the proximal GATA recognition motif in the P1 promoter and impact on the maturation of furin substrates. Blood **100**:3578–3587

[31] Nagahama M, Taniguchi T, Hashimoto E, Imamaki A, Mori K, Tsuji A, Matsuda Y (1998) Biosynthetic processing and quaternary interactions of proprotein convertase SPC4 (PACE4). FEBS Lett **434**:155–159

[32] Taniguchi T, Kuroda R, Sakurai K, Nagahama M, Wada I, Tsuji A, Matsuda Y (2002) A critical role for the carboxy terminal region of the proprotein convertase, PACE4A, in the regulation of its autocatalytic activation coupled with secretion. Biochem Biophys Res Commun **290**:878–884

[33] Tsuji A, Sakurai K, Kiyokage E, Yamazaki T, Koide S, Toida K, Ishimura K, Matsuda Y (2003) Secretory proprotein convertases PACE4 and PC6A are heparin-binding proteins which are localized in the extracellular matrix. Potential role of PACE4 in the activation of proproteins in the extracellular matrix. Biochim Biophys Acta **1645**:95–104

[34] Nour N, Mayer G, Mort JS, Salvas A, Mbikay M, Morrison CJ, Overall CM, Seidah NG (2005) The cysteine-rich domain of the secreted proprotein convertases PC5A and PACE4 functions as a cell surface anchor and interacts with tissue inhibitors of metalloproteinases. Mol Biol Cell **16**:5215–5226

[35] Bando M, Matsuoka A, Tsuji A, Matsuda Y (2002) The proprotein convertase PACE4 is upregulated by PDGF-BB in megakaryocytes: Gene expression of PACE4 and furin is regulated differently in Dami cells. J Biochem (Tokyo) **132**:127–134

[36] Blanchette F, Day R, Dong W, Laprise MH, Dubois CM (1997) TGFbeta1 regulates gene expression of its own converting enzyme furin. J Clin Invest **99**:1974–1983

[37] Yoshida I, Koide S, Hasegawa SI, Nakagawara A, Tsuji A, Matsuda Y (2001) Proprotein convertase PACE4 is down-regulated by the basic helix-loop-helix transcription factor hASH-1 and MASH-1. Biochem J **360**:683–689

[38] Nakayama K, Kim WS, Torii S, Hosaka M, Nakagawa T, Ikemizu J, Baba T, Murakami K (1992) Identification of the fourth member of the mammalian endoprotease family homologous to the yeast Kex2 protease. Its testis-specific expression. J Biol Chem **267**:5897–5900

[39] Seidah NG, Day R, Hamelin J, Gaspar A, Collard MW, Chretien M (1992) Testicular expression of PC4 in the rat: Molecular diversity of a novel germ cell-specific Kex2/subtilisin-like proprotein convertase. Mol Endocrinol **6**:1559–1570

[40] Torii S, Yamagishi T, Murakami K, Nakayama K (1993) Localization of Kex2-like processing endoproteases, furin and PC4, within mouse testis by in situ hybridization. FEBS Lett **316**:12–16

[41] Li M, Mbikay M, Nakayama K, Miyata A, Arimura A (2000) Prohormone convertase PC4 processes the precursor of PACAP in the testis. Ann N Y Acad Sci **921**:333–339

[42] Lusson J, Vieau D, Hamelin J, Day R, Chretien M, Seidah NG (1993) cDNA structure of the mouse and rat subtilisin/kexin-like PC5: A candidate proprotein convertase expressed in endocrine and nonendocrine cells. Proc Natl Acad Sci U S A **90**:6691–6695

[43] Nakagawa T, Hosaka M, Torii S, Watanabe T, Murakami K, Nakayama K (1993) Identification and functional expression of a new member of the mammalian Kex2-like processing endoprotease family: Its striking structural similarity to PACE4. J Biochem (Tokyo) **113**:132–135

[44] Stawowy P, Blaschke F, Kilimnik A, Goetze S, Kallisch H, Chretien M, Marcinkiewicz M, Fleck E, Graf K (2002) Proprotein convertase PC5 regulation by PDGF-BB involves PI3-kinase/p70(s6)-kinase activation in vascular smooth muscle cells. Hypertension **39**:399–404

[45] Siegfried G, Khatib AM, Benjannet S, Chretien M, Seidah NG (2003) The proteolytic processing of pro-platelet-derived growth factor-A at RRKR(86) by members of the proprotein convertase family is functionally correlated to platelet-derived growth factor-A-induced functions and tumorigenicity. Cancer Res **63**:1458–1463

[46] Siegfried G, Basak A, Prichett-Pejic W, Scamuffa N, Ma L, Benjannet S, Veinot JP, Calvo F, Seidah N, Khatib AM (2005) Regulation of the stepwise proteolytic cleavage and secretion of PDGF-B by the proprotein convertases. Oncogene **24**:6925–6935

[47] Siegfried G, Basak A, Cromlish JA, Benjannet S, Marcinkiewicz J, Chretien M, Seidah NG, Khatib AM (2003) The secretory proprotein convertases furin, PC5, and PC7 activate VEGF-C to induce tumorigenesis. J Clin Invest **111**:1723–1732

[48] Lissitzky JC, Luis J, Munzer JS, Benjannet S, Parat F, Chretien M, Marvaldi J, Seidah NG (2000) Endoproteolytic processing of integrin pro-alpha subunits involves the redundant function of furin

and proprotein convertase (PC) 5A, but not paired basic amino acid converting enzyme (PACE) 4, PC5B or PC7. Biochem J **346**:133–138

[49] Kalus I, Schnegelsberg B, Seidah NG, Kleene R, Schachner M (2003) The proprotein convertase PC5A and a metalloprotease are involved in the proteolytic processing of the neural adhesion molecule L1. J Biol Chem **278**:10381–10388

[50] Ulloa L, Creemers JW, Roy S, Liu S, Mason J, Tabibzadeh S (2001) Lefty proteins exhibit unique processing and activate the MAPK pathway. J Biol Chem **276**:21387–21396

[51] Nour N, Basak A, Chretien M, Seidah NG (2003) Structure-function analysis of the prosegment of the proprotein convertase PC5A. J Biol Chem **278**:2886–2895

[52] Seidah NG, Hamelin J, Mamarbachi M, Dong W, Tardos H, Mbikay M, Chretien M, Day R (1996) cDNA structure, tissue distribution, and chromosomal localization of rat PC7, a novel mammalian proprotein convertase closest to yeast kexin-like proteinases. Proc Natl Acad Sci U S A **93**:3388–3393

[53] Bruzzaniti A, Goodge K, Jay P, Taviaux SA, Lam MH, Berta P, Martin TJ, Moseley JM, Gillespie MT (1996) PC8 [corrected], a new member of the convertase family. Biochem J **314**:727–731

[54] Meerabux J, Yaspo ML, Roebroek AJ, Van de Ven WJ, Lister TA, Young BD (1996) A new member of the proprotein convertase gene family (LPC) is located at a chromosome translocation breakpoint in lymphomas. Cancer Res **56**:448–451

[55] Taylor NA, Van De Ven WJ, Creemers JW (2003) Curbing activation: Proprotein convertases in homeostasis and pathology. FASEB J **17**:1215–1227

[56] Constam DB, Robertson EJ (1999) Regulation of bone morphogenetic protein activity by pro domains and proprotein convertases. J Cell Biol **144**:139–149

[57] Constam DB, Calfon M, Robertson EJ (1996) SPC4, SPC6, and the novel protease SPC7 are coexpressed with bone morphogenetic proteins at distinct sites during embryogenesis. J Cell Biol **134**:181–191

[58] Seidah NG, Mowla SJ, Hamelin J, Mamarbachi AM, Benjannet S, Toure BB, Basak A, Munzer JS, Marcinkiewicz J, Zhong M, Barale JC, Lazure C, Murphy RA, Chretien M, Marcinkiewicz M (1999) Mammalian subtilisin/kexin isozyme SKI-1: A widely expressed proprotein convertase with a unique cleavage specificity and cellular localization. Proc Natl Acad Sci U S A **96**:1321–1326

[59] Sakai J, Rawson RB, Espenshade PJ, Cheng D, Seegmiller AC, Goldstein JL, Brown MS (1998) Molecular identification of the sterol-regulated luminal protease that cleaves SREBPs and controls lipid composition of animal cells. Mol Cell **2**:505–514

[60] Brown MS, Goldstein JL, (1997) The SREBP pathway: Regulation of cholesterol metabolism by proteolysis of a membrane-bound transcription factor. Cell **89**:331–340

[61] Lenz O, ter Meulen J, Klenk HD, Seidah NG, Garten W (2001) The Lassa virus glycoprotein precursor GP-C is proteolytically processed by subtilase SKI-1/S1P. Proc Natl Acad Sci U S A **98**:12701–12705

[62] Ye J, Rawson RB, Komuro R, Chen X, Dave UP, Prywes R, Brown MS, Goldstein JL (2000) ER stress induces cleavage of membrane-bound ATF6 by the same proteases that process SREBPs. Mol Cell **6**:1355–1364

[63] Yang J, Goldstein JL, Hammer RE, Moon YA, Brown MS, Horton JD (2001) Decreased lipid synthesis in livers of mice with disrupted Site-1 protease gene. Proc Natl Acad Sci U S A **98**:13607–13612

[64] Mouchantaf R, Watt HL, Sulea T, Seidah NG, Alturaihi H, Patel YC, Kumar U (2004) Prosomatostatin is proteolytically processed at the amino terminal segment by subtilase SKI-1. Regul Pept **120**:133–140

[65] Stirling J, O'hare P (2006) CREB4, a Transmembrane bZip Transcription Factor and Potential New Substrate for Regulation and Cleavage by S1P. Mol Biol Cell **17**:413–426

[66] Seidah NG, Benjannet S, Wickham L, Marcinkiewicz J, Jasmin SB, Stifani S, Basak A, Prat A, Chretien M (2003) The secretory proprotein convertase neural apoptosis-regulated convertase 1 (NARC-1): Liver regeneration and neuronal differentiation. Proc Natl Acad Sci U S A **100**:928–933

[67] Naureckiene S, Ma L, Sreekumar K, Purandare U, Frederick Lo C, Huang Y, Chiang LW, Grenier JM, Ozenberger BA, Steven Jacobsen J, Kennedy JD, DiStefano PS, Wood A, Bingham B (2003)

Functional characterization of Narc 1, a novel proteinase related to proteinase K. Arch Biochem Biophys **420**:55–67

[68] Seidah NG (2001) Cellular limited proteolysis of precursor proteins and peptides. In: Dalbey RE, Sigman DS (eds), The Enzymes: Co- and Posttranslational Proteolysis of Proteins, San Diego, CA, Academic Press, pp 237–258

[69] Abifadel M, Varret M, Rabes JP, Allard D, Ouguerram K, Devillers M, Cruaud C, Benjannet S, Wickham L, Erlich D, Derre A, Villeger L, Farnier M, Beucler I, Bruckert E, Chambaz J, Chanu B, Lecerf JM, Luc G, Moulin P, Weissenbach J, Prat A, Krempf M, Junien C, Seidah NG, Boileau C (2003) Mutations in PCSK9 cause autosomal dominant hypercholesterolemia. Nat Genet **34**:154–156

[70] Attie AD, Seidah NG (2005) Dual regulation of the LDL receptor – some clarity and new questions. Cell Metab **1**:290–292

[71] Cohen J, Pertsemlidis A, Kotowski IK, Graham R, Garcia CK, Hobbs HH (2005) Low LDL cholesterol in individuals of African descent resulting from frequent nonsense mutations in PCSK9. Nat Genet **37**:161–165

[72] Jutras I, Seidah NG, Reudelhuber TL, Brechler V (1997) Two activation states of the prohormone convertase PC1 in the secretory pathway. J Biol Chem **272**:15184–15188

[73] Fricker LD, McKinzie AA, Sun J, Curran E, Qian Y, Yan L, Patterson SD, Courchesne PL, Richards B, Levin N, Mzhavia N, Devi LA, and Douglass J (2000) Identification and characterization of proSAAS, a granin-like neuroendocrine peptide precursor that inhibits prohormone processing. J. Neurosci **20**:639–648

[74] Basak A, Koch P, Dupelle M, Fricker LD, Devi LA, Chretien M, and Seidah NG (2001) Inhibitory specificity and potency of proSAAS-derived peptides toward proprotein convertase 1. J. Biol. Chem **276**:32720–32728

[75] Mbikay M, Seidah NG, and Chretien M (2001) Neuroendocrine secretory protein 7B2: Structure, expression and functions. Biochem. J **357**:329–342

[76] Hsi KL, Seidah NG, De Serres G, Chretien M (1982) Isolation and NH2-terminal sequence of a novel porcine anterior pituitary polypeptide. Homology to proinsulin, secretin and Rous sarcoma virus transforming protein TVFV60. FEBS Lett. **147**:261–266

[77] Seidah NG, Hsi KL, De Serres G, Rochemont J, Hamelin J, Antakly T, Cantin M, Chretien M (1983) Isolation and NH2-terminal sequence of a highly conserved human and porcine pituitary protein belonging to a new superfamily. Immunocytochemical localization in pars distalis and pars nervosa of the pituitary and in the supraoptic nucleus of the hypothalamus. Arch Biochem Biophys **225**:525–534

[78] Marcinkiewicz M, Benjannet S, Falgueyret JP, Seidah NG, Schurch W, Verdy M, Cantin M, Chretien M (1988) Identification and localization of 7B2 protein in human, porcine, and rat thyroid gland and in human medullary carcinoma. Endocrinology **123**:866–873

[79] Fricker LD, McKinzie AA, Sun J, Curran E, Qian Y, Yan L, Patterson SD, Courchesne PL, Richards B, Levin N, Mzhavia N, Devi LA, Douglass J (2000) Identification and characterization of proSAAS, a granin-like neuroendocrine peptide precursor that inhibits prohormone processing. J Neurosci **20**:639–648

[80] Naggert JK, Fricker LD, Varlamov O, Nishina PM, Rouille Y, Steiner DF, Carroll RJ, Paigen BJ, Leiter EH (1995) Hyperproinsulinaemia in obese fat/fat mice associated with a carboxypeptidase E mutation which reduces enzyme activity. Nat Genet **10**:135–142

[81] Qian Y, Devi LA, Mzhavia N, Munzer S, Seidah NG, Fricker LD (2000) The C-terminal region of proSAAS is a potent inhibitor of prohormone convertase 1. J Biol Chem **275**:23596–23601

[82] Zhu, X, Lindberg I, (1995) 7B2 facilitates the maturation of proPC2 in neuroendocrine cells and is required for the expression of enzymatic activity. J Cell Biol **129**:1641–1650

[83] Benjannet S, Rondeau N, Day R, Chretien M, Seidah N (1991) PC1 and PC2 are proprotein convertases capable of cleaving proopiomelanocortin at distinct pairs of basic residues. Proc. Natl. Acad. Sci. U. S. A **88**:3564–3568

[84] Boudreault A, Gauthier D, Rondeau N, Savaria D, Seidah N, Chretien M, Lazure C (1998) Molecular characterization, enzymatic analysis, and purification of murine proprotein convertase-1/3 (PC1/PC3) secreted from recombinant baculovirus-infected insect cells. Prot. Exp. Purif **14**:353–366

[85] Wei S, Feng Y, Che FY, Pan H, Mzhavia N, Devi LA, McKinzie AA, Levin N, Richards WG, Fricker LD (2004) Obesity and diabetes in transgenic mice expressing proSAAS. J Endocrinol **180**:357–368

[86] Anderson ED, VanSlyke JK, Thulin CD, Jean F, Thomas G (1997) Activation of the furin endoprotease is a multiple-step process: Requirements for acidification and internal propeptide cleavage. EMBO J **16**:1508–1518

[87] Zhong M, Munzer JS, Basak A, Benjannet S, Mowla SJ, Decroly E, Chrétien M, Seidah NG (1999) The prosegments of furin and PC7 as potent inhibitors of proprotein convertases: In vitro and ex vivo assessment of their specificity and selectivity. J. Biol. Chem **274**:33913–33920

[88] Bhattacharjya S, Xu P, Zhong M, Chretien M, Seidah NG, Ni F (2000) Inhibitory activity and structural characterization of a C-terminal peptide fragment derived from the prosegment of the proprotein convertase PC7. Biochemistry **39**:2868–2877

[89] Boudreault A, Gauthier D, Lazure C (1998) Proprotein convertase PC1/3-related peptides are potent slow tight-binding inhibitors of murine PC1/3 and Hfurin. J Biol Chem **273**:31574–31580

[90] Lazure C, Gauthier D, Jean F, Boudreault A, Seidah NG, Bennett HP, Hendy GN (1998) In vitro cleavage of internally quenched fluorogenic human proparathyroid hormone and proparathyroid-related peptide substrates by furin. Generation of a potent inhibitor. J Biol Chem **273**:8572–8580

[91] Sawada Y, Inoue M, Kanda T, Sakamaki T, Tanaka S, Minamino N, Nagai R, Takeuchi T (1997) Co-elevation of brain natriuretic peptide and proprotein-processing endoprotease furin after myocardial infarction in rats. FEBS Lett. **400**:177–182

[92] Lu W, Zhang W, Molloy SS, Thomas G, Ryan K, Chiang Y, Anderson S, Laskowski Jr M (1993) Arg15-Lys17-Arg18 turkey ovomucoid third domain inhibits human furin. J Biol Chem **268**:14583–14585

[93] Garten W, Hallenberger S, Ortmann D, Schafer W, Vey M, Angliker H, Shaw E, Klenk HD (1994) Processing of viral glycoproteins by the subtilisin-like endoprotease furin and its inhibition by specific peptidylchloroalkylketones. Biochimie **76**:217–225

[94] Bathurst IC, Brennan SO, Carrell RW, Cousens LS, Brake AJ, Barr PJ (1987) Yeast KEX2 protease has the properties of a human proalbumin converting enzyme. Science **235**:348–350

[95] Brennan SO, Peach RJ (1988) Calcium-dependent KEX2-like protease found in hepatic secretory vesicles converts proalbumin to albumin. FEBS Lett. **229**:167–170

[96] Anderson ED, Thomas L, Hayflick JS, Thomas G (1993) Inhibition of HIV-1 gp160-dependent membrane fusion by a furin-directed α 1-antitrypsin variant. J Biol Chem **268**:24887–24891

[97] Jean F, Stella K, Thomas L, Liu G, Xiang Y, Reason AJ, Thomas G (1998) α1-Antitrypsin Portland, a bioengineered serpin highly selective for furin: Application as an antipathogenic agent. Proc Natl Acad Sci USA **23**:7293–7298

[98] Dufour EK, Denault JB, Hopkins PC, Leduc R (1998) Serpin-like properties of alpha1-antitrypsin Portland towards furin convertase. FEBS Lett. **426**:41–46

[99] Vollenweider F, Benjannet S, Decroly E, Savaria D, Lazure C, Thomas G, Chrétien M and Seidah NG (1996) Comparative cellular processing of the human immunodeficiency virus (HIV-1) envelope glycoprotein gp160 by the mammalian subtilisin/kexin-like convertases. Biochem J **314**:521–532

[100] Benjannet S, Savaria D, Laslop A, Munzer JS, Chretien M, Marcinkiewicz M, Seidah NG (1997) Alpha1-antitrypsin Portland inhibits processing of precursors mediated by proprotein convertases primarily within the constitutive secretory pathway **272**:26210–26218

[101] Munzer JS, Basak A, Zhong M, Mamarbachi M, Hamelin J, Savaria D, Lazure C, Benjannet S, Chrétien M and Seidah NG (1997) In Vitro characterization of the novel proprotein convertase PC7. J Biol Chem **272**:19672–19681

[102] Lopez-Perez E, Seidah NG, Checler F (1999) Proprotein convertase activity contributes to the processing of the Alzheimer's β-amyloid precursor protein in human cells: Evidence for a role of the prohormone convertase PC7 in the constitutive α-secretase pathway. J Neurochem **73**:2056–2062

[103] Abrami L, Fivaz M, Decroly E, Seidah NG, Jean F, Thomas G, Leppla SH, Buckley JT, van der Goot FG, (1998) The pore-forming toxin proaerolysin is activated by furin. J Biol Chem **273**:32656–32661

[104] Cui Y, Jean F, Thomas G, Christian JL (1998) BMP-4 is proteolytically activated by furin and/or PC6 during vertebrate embryonic development. EMBO J **17**:4735–4743

[105] Khatib AM, Siegfried G, Prat A, Luis J, Chretien M, Metrakos P, Seidah NG (2001) Inhibition of proprotein convertases is associated with loss of growth and tumorigenicity of HT-29 human colon carcinoma cells: Importance of insulin-like growth factor-1 (IGF-1) receptor processing in IGF-1-mediated functions. J Biol Chem **276**:30686–30693

[106] Van Rompaey L, Ayoubi T, Van De Ven W, Marynen P (1997) Inhibition of intracellular proteolytic processing of soluble proproteins by an engineered alpha 2-macroglobulin containing a furin recognition sequence in the bait region. Biochem J **326**:507–514

[107] Dahlen JR, Jean F, Thomas G, Foster C, Kisiel W (1998) Inhibition of Soluble Recombinant Furin by Human Proteinase Inhibitor 8. J Biol Chem **273**:1851–1854

[108] Cameron A, Appel J, Houghten RA, Lindberg I (2000) Polyarginines Are Potent Furin Inhibitors. J Biol Chem **275**:36741–36749

[109] Roebroek AJ, Umans L, Pauli IG, Robertson EJ, van Leuven F, Van de Ven WJ, Constam DB (1998) Failure of ventral closure and axial rotation in embryos lacking the proprotein convertase furin. Development **125**:4863–4876

[110] Essalmani R, Hamelin J, Marcinkiewwicz E, Chamberland A, Mbikay M, Chretien C, Seidah N.G, Prat A (2006) Genetic deletion of PC/6 leads to early embryonic lethalithy. Mol.Cell Biol **26**:354–361

[111] Yang J, Goldstein JL, Hammer RE, Moon YA, Brown MS, Horton JD (2001) Decreased lipid synthesis in livers of mice with disrupted Site-1 protease gene. Proc Natl Acad Sci U S A **98**:13607–13612

[112] Furuta M, Yano H, Zhou A, Rouille Y, Holst JJ, Carroll R, Ravazzola M, Orci L, Furuta H, Steiner DF (1997) Defective prohormone processing and altered pancreatic islet morphology in mice lacking active SPC2. Proc Natl Acad Sci U S A **94**:6646–6651

[113] Zhu X, Zhou A, Dey A, Norrbom C, Carroll R, Zhang C, Laurent V, Lindberg I, Ugleholdt R, Holst JJ, Steiner DF (2002) Disruption of PC1/3 expression in mice causes dwarfism and multiple neuroendocrine peptide processing defects. Proc Natl Acad Sci U S A **99**:10293–10298

[114] Constam DB, Robertson EJ (2000) SPC4/PACE4 regulates a TGFbeta signaling network during axis formation. Genes Dev. **14**:1146–1155

[115] Mbikay M, Tadros H, Ishida N, Lerner CP, De Lamirande E, Chen A, El-Alfy M, Clermont Y, Seidah NG, Chretien M, Gagnon C, Simpson EM (2002) Disruption of PC1/3 expression in mice causes dwarfism and multiple neuroendocrine peptide processing defects. Proc Natl Acad Sci U S A **99**:10293–10298

[116] Milner JM, Rowan AD, Elliott SF, Cawston TE (2003) Inhibition of furin-like enzymes blocks interleukin-1alpha/oncostatin M-stimulated cartilage degradation. Arthritis Rheum **48**:1057–1066

CHAPTER 2

SIGNALLING PATHWAYS LEADING TO FURIN EXPRESSION IN CANCER

STEPHANIE McMAHON AND CLAIRE M. DUBOIS

Immunology Division, Faculty of Medicine, Université de Sherbrooke, Sherbrooke, Qc, Canada, J1H 5N4

Abstract: With the continually increasing body of evidence implicating furin in health as well as in a large number of diseases, it has become important to identify the molecules involved in the signaling pathways leading to gene expression of this enzyme. Furin is dramatically overexpressed in cancer and various inflammatory conditions clearly pointing to the molecular mechanisms controlling its expression as a potential target for eventual rational therapeutic intervention. Currently, regulation of the most proximal furin promoter is beginning to be understood in the context of activation of certain signalling pathways involved in the growth and progression of neoplasms. In addition, several factors that modulate the furin-transcriptional unit are being elucidated at the level of cis-acting sequences. Strategies that take advantage of the signalling pathways that regulate furin expression for the treatment of cancer will be discussed

Keywords: Proprotein convertases, furin, HIF-1, MAPKs, Smads, TGFβ, hypoxia

1.   INTRODUCTION

Tumor growth and malignant tumor phenotypes are regulated by the action of proprotein convertases (PCs), especially furin [1]. In normal tissues, furin is typically detected at very low levels; however, elevated expression of this convertase has been reported in a variety of human cancers. Overexpression of furin mRNA was detected in breast tumors, head and neck tumors, and lung cancer using RT-PCR and *in situ* hybridization techniques [2–4]. Furin protein was also demonstrated to be expressed at high levels in human gliomas by immunochemistry, and in head and neck tumors by Western blotting [4, 5]. Moreover, furin expression has been correlated with cancer aggressiveness and was therefore proposed to have significant prognostic value [4, 6]. This suggests that the potentially deleterious effects of

A-Majid Khatib (ed.), Regulation of Carcinogenesis, Angiogenesis and Metastasis by the Proprotein Convertases, 27–45.
© 2006 *Springer.*

furin on the maintenance of cellular homeostasis under physiological conditions are avoided by very low cellular levels of expression.

2.   IMPACT OF TUMOR MICROENVIRONMENT ON FURIN EXPRESSION

Over the recent years, a better understanding of the influence of the tumor micro-environment on furin expression has emerged. Until now, only few cancer related molecules were shown to modulate furin expression. Among others, the parathyroid hormone-related peptide (PTHrP), known to regulate the growth and invasion of human cancers such as breast, prostate, and lung cancer, enhances furin mRNA expression when added to human gastric cancer cell cultures. The transfection of furin cDNA within these cells markedly increased the production of mature PTHrP, a natural furin substrate, as well as cell proliferation, suggesting a link between furin expression, PHTrP maturation and gastric cancer cell growth [7]. Moreover, BCL-2, a pro-apoptotic molecule overexpressed in several cancers known to enhance migration and invasion of glioma cells, was also shown to have an impact on furin expression. In fact, BCL-2 glioma-expressing cells exhibit enhanced expression and activity of the proprotein convertase furin, as well as enhanced MMP activity [8]. However, little is known about the molecular mechanisms by which the *fur* gene, encoding furin, is differentially expressed and regulated by these mediators. More recently, common tumorigenesis enhancers, TGFβ1 and hypoxia, which are typically present in solid tumors, were uncovered as regulators of furin expression and some of the proteins involved in their signalling pathway were identified.

3.   FURIN REGULATION BY TGFβ

It has become apparent over the past few years that transforming growth factor beta (TGFβ) is the cytokine that possesses one of the most significant and versatile role in cancer, being involved in its susceptibility, development and progression [9–11]. It is a potent inhibitor of normal stromal, hematopoietic, and epithelial cell growth. However, in the later stages of cancer development TGFβ is actively secreted by tumor cells where it does not act as a bystander but contributes to cell growth, invasion, metastasis, and decreases in host-tumor immune responses. At these stages, alterations in the TGFβ responsiveness or bioac-tivation pathways have been observed in human cancer cell lines and/or neoplasms. This includes mutation or deletion of members of the signalling pathway, resis-tance to TGFβ-mediated inhibition of proliferation, or mutations in its converting enzyme, furin. Identifying and understanding the TGFβ production/activation pathway abnormalities present in various malignancies is currently considered as a promising avenue of study that may yield new modalities to both prevent and treat cancer.

4. PROTEOLYTIC PROCESSING OF TGFβ BY FURIN

A close relationship has been observed between furin and TGFβ that applies to cancer development. Most cell types secrete large quantities of TGFβ as a latent, inactive complex. TGFβ is first synthesized as an inactive precursor that dimerizes and subsequently requires two processing reactions for activation. The first step is an enzymatic cleavage at a basic sequence to yield the amino-terminal latency-associated peptide (LAP) and carboxyl-terminal peptide (TGFβ) [12, 13]. This maturation step occurs in the secretory pathway and is known to be mediated by PC family members, in particular by furin [14]. The LAP portion remains associated with TGFβ via electrostatic interactions, thus forming the secreted latent TGFβ. LAP is often linked through disulfide bonding to the latent TGFβ binding protein (LTBP) allowing deposition of the complex to the extracellular cell matrix [12, 13]. As a second reaction, LAP must be released from the latent complex before TGF-β can bind and activate its receptors. A variety of agents activate latent TGF-β such as conformational changes triggered by thrombospondin or the integrin αvβ6 and/or cleavage by plasmin or members of the matrix metalloprotease family (Figure 1)

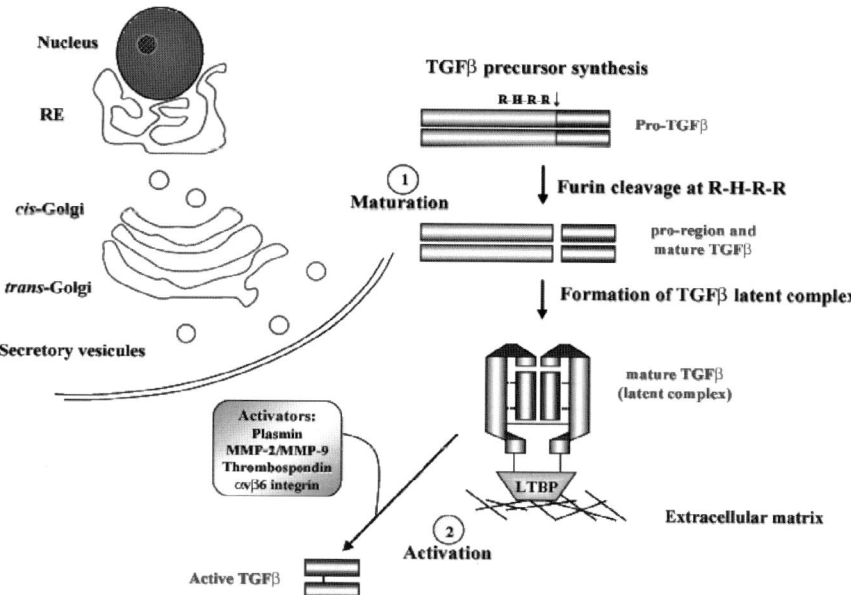

Figure 1. Schematic representation of the maturation and activation of TGFβ1 precursor. The bioactive form of TGFβ1 is obtained through the removal of the pro-domain by furin proteolytic processing following the sequence R-H-R-R (maturation step) within the *trans*-Golgi network. Mature TGFβ1 remains non-covalently linked to its pro-region to form the latent complex which is secreted by cells. This complex can bind to the latent TGFβ-binding protein (LTBP) that permits its anchorage and storage within the extracellular matrix. The dissociation of the pro-segment (activation step), which can be mediated by several molecules that act through direct cleavage of the pro-region or conformational changes, induces the liberation of the active TGFβ1 dimer

[15, 16]. Yet, regardless of the type of reaction involved, furin processing is always required and was shown to be an essential step leading to TGF-β bioactivation [14, 17, 18]. This is illustrated by studies indicating that LoVo cells, an adenocarcinoma cell line in which furin is invalidated through mutations in both alleles, do not produce significant amounts of bioactivable TGF-β, a defect that can be reconstituted by expression of the wild type *fur* gene [14]. In addition, the expression of the furin inhibitor α_1-PDX in various cell lines, including the hepatoma cell line Hepa-1c1c7 and the kidney cells line BSC-40, resulted in the secretion of the intact TGFβ precursor form resistant to heat-induced activation [14, 19].

5.   TGFβ AUTOREGULATORY LOOP

One interesting particularity of the TGFβ system is its ability to regulate the effectiveness of several important keypoints in its production/activation pathway. For example, TGFβ modulates the transcription of the plasminogen activator/plasminogen activator-inhibitor system involved in the dissociation of the latent TGFβ complex [20] and is a potent inducer of its own expression through the induction of the Smad-AP-1 complex [21, 22]. The expression of the *fur* gene was also shown to be transcriptionally upregulated by TGFβ1 in various normal cells as well as in hepatocyte carcinoma HepG2 [23, 24] generating an unique enzyme/substrate amplification loop that leads to an increase in local concentrations of TGFβ. Such an adaptative responsiveness of the TGFβ convertase has fundamental implications in TGFβ-related pathologies. It has encouraged work dedicated to the understanding of the mechanisms that control its expression.

6.   ROLE OF THE SMAD PATHWAY IN FURIN GENE
REGULATION

In the last few years, significant progress has been made in elucidating the signalling mechanisms utilized by TGFβ. A major discovery comes from the recent uncovering and cloning of specific Smad signalling proteins that consist in pathway-restricted Smads (Smad2 and Smad3), the common mediator Smad4, and the inhibitory Smads (Smad6 and Smad7) [25–27] that directly inhibit the TGFβ type I receptor serine-threonine kinase and its transcriptional effectors. TGFβ signals through sequential activation of two cell-surface serine-threonine kinase receptors, which phosphorylate Smad2 and Smad3 [28, 29]. These activated Smads, together with Smad4, translocate to the nucleus and, in association with other transcription factors, activate the transcription of target genes [30, 31]. Among the participating transcription factors are the winged helix factor FAST (now known as FoxH1) [32] in the case of Smad2 [33] or AP-1 in the case of Smad3 [22, 34, 35]. Naturally occurring inhibitors, Smad6 and Smad7, block and hence control TGFβ superfamily signalling by preventing phosphorylation and/or inducing dephosphorylaion of receptor-associated Smads and/or by inducing receptor complex degradation through the recruitment of ubiquitin-ligases that induce proteasomal degradation [36].

In a recent study, evidence was provided for the central role of Smads in the transcriptional activation of the TGFβ-inducible *fur* promoter [23]. As described by Ayoubi et al., three alternative promoters direct *fur* expression [37]. Promoters P1A and P1B are very GC-rich and contain several SP1 sites, so they have been proposed to be involved in maintaining the housekeeping levels of this enzyme. Promoter P1, on the other hand, is of the inducible type. It has both TATA and CCAAT elements in the proximal promoter region, and was reported to be *trans*-activated by the transcription factor C/EBP, a feature relevant to its relatively high levels of expression in hepatocytes. Using HepG2 cells transfected with various furin promoter constructs, it was observed that among the three furin promoters, the P1 promoter was the strongest and most sensitive to TGFβ1 and that Smad2 together with winged helix transcription factor FAST (FoxH1) participated in this transactivation [23]. It was also found that the proximal P1 promoter region, that contains one SBE (Smad binding element) and one ARE (activin-responsive element) binding site, carried most of the Smad responsiveness. Another member of the Smad pathway, namely Smad3 in association with Smad4 also played a role in the production of furin (Blanchette F. et al., unpublished observation). Furthermore, subsequent studies have showed that other kinases such as the mitogen-activated protein kinase (MAPK) can participate in the signalling pathway leading to furin expression.

7. ROLE OF THE RAS-ERK PATHWAY IN FURIN GENE REGULATION

Ras proteins are small G-proteins with GTPase activity that are located at the cytoplasmic membrane [38]. They are activated downstream of a variety of trans-membrane receptors, a process that involves the exchange of GTP for GDP. Activated Ras represents an important signalling branchpoint because they activate several signalling pathways through a number of effectors that serve to regulate myriad of cell functions including growth, survival, differentiation, and angio-genesis. As important effectors are the protein kinases of the Raf family that act upstream to a two-tiered protein kinase cascade, which includes two other cytosolic protein kinases, MEK and ERK. ERK phosphorylates many substrates involved in diverse cellular functions and was shown to play an important role in determining cell fate, selecting between diverse responses such as proliferation, differentiation, senescence or survival. Using HepG2 cells transfected with the furin P1 promoter construct, it was observed that forced expression of a dominant negative mutant form of Ras (RasN17) inhibited TGF beta 1-induced furin gene transcription, suggesting the involvement of Ras in this regulation [39]. In additional studies, the use of antisense and pharmacological inhibitors of ERK has demonstrated that the ERK signalling pathway acts downstream of Ras in furin gene regulation expression [39]. In this context, dysregulation of Ras proteins by activating mutations, overex-pression or upstream activation was observed in approximately 30% of all human tumors, a feature that may explain, in part, the increased levels of this convertase found in a wide array of cancer types. Since the activated Ras-MAPK pathway is coupled to growth factor receptors, including EGF or PDGF and favor cellular proliferation, differentiation and survival, convertase activation of growth factor and/or receptors which signal through Ras, would likely amplify the Ras-MAPK signalling pathway.

In light of the emerging evidence for the interactions between the Smad and MAPK pathways and the role of furin in growth and differentiation events, it was of interest to explore the possible integration of these two pathways for the regulation of this convertase. Using HepG2 cells, it was observed that inhibition of MEK by PD98059 blocked most of the enhanced Smad2 nuclear localization induced by TGFβ. In contrast, activation of ERK1/2 by activated MEK1 resulted in an enhanced nuclear localization of Smad2 [39]. These findings suggest functional interactions between the Smad and the MEK/ERK pathway in furin regulation. The cross-talk between these two signalling pathway may serve a growth/differentiation integration signal for the bioavailability of a multitude of furin-activated precursors involved in tumor growth and metastasis (Figure 2).

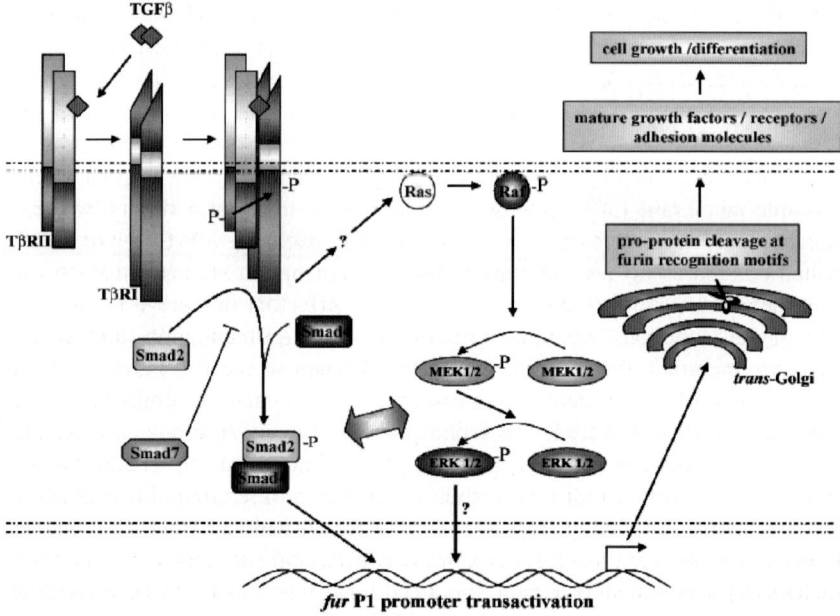

Figure 2. Cooperative model for TGFβ-activated fur gene expression. The addition of TGFβ1 to cells results in the activation of the TGFβ-specific Smad2/4 pathway as shown. In parallel, TβRII/I activation induces a rapid and sustained phosphorylation of endogenous ERK1/2 MAPK. Cross-talk with activated MEK 1/2 or downstream MAPK cascade elements enhances Smad2 nuclear translocation where it may interact with DNA-binding proteins and direct transcription of the *fur* gene. The increase of furin intracellular levels will impact the bioactivation of multiple growth/cell differentiation-related factors

In addition to members of the vast TGFβ family, these precursor proteins include, among others, several key growth factor precursors such as platelet-derived growth factor A and B chains, growth factor proreceptors such as the insulin receptor and the hepatocyte growth factor receptor (c-Met), several integrin α-subunits, and cadherin family members, many of them were found to activate RAS-MAPK.

The outcome of Smad-MAPK interactions for enhanced furin expression should therefore not be viewed solely as a result of multifaceted TGFβ signalling downstream of its receptor but also as a consequence of growth factors/receptors network acting together to amplify Smad/MAPK signal.

8.   FURIN REGULATION BY HYPOXIA

8.1   Tumor Hypoxia

The tumor progression process is characterized by rapid cell growth accompanied by alterations in the microenvironment of the tumor cells. Among these changes, is the appearance of multiple areas of hypoxia (low oxygen tension), a hallmark feature of human and experimental tumors [40, 41]. Hypoxic regions within tumors arise for several reasons. First of all, rapid tumor cell growth can outpace the distance that allows adequate delivery of O_2 by surrounding blood vessels, causing hypoxic areas to form. Secondly, new blood vessels formed in tumors are usually disorganized and are prone to collapse, resulting in regions of inadequate perfusion and hypoxia. Additionally, diminished tumor oxygenation can be exarcerbated by reduced O_2 transport caused by cancer-related or cancer treatment-induced anemia [42].

Hypoxia has recently emerged as a major factor that influences malignant progression. The effects of hypoxia are mainly mediated by a series of hypoxia-induced proteomic and genomic changes that enable tumor cells to survive or escape their oxygen deficient environment [43, 44]. But how does hypoxia contribute to tumor progression? One of the key factors in mediating tumor cell survival and tumor progression is the hypoxically regulated production of growth factors that induce new blood vessel formation (angiogenesis). In tumors, the ability to induce angiogenesis, in addition to provide O_2 and nutrients to support cell survival and proliferation, is associated with the development of metastasis by increasing the opportunity of tumoral cells to access blood circulation and disseminate at distal sites [45, 46].

Hypoxia-inducible factor 1 (HIF-1) is one of the master regulators that orchestrate the cellular response to hypoxia [47, 48]. This transcription factor is a heterodimer composed of two subunits: an oxygen sensitive alpha-subunit (HIF-1α) and a constitutively expressed beta-subunit (ARNT/HIF-1β) [49, 50]. Under normoxic conditions, HIF-1α is hydroxylated on several specific proline residues by one of three HIF-1 oxygen-dependant prolyl hydroxylases, PHD1-3, a reaction that depends on the presence of oxygen. Hydroxylation permits the binding of HIF-1α to the von Hippel Lindau tumor suppressor (VHL), a component of an E3 ubiquitin ligase complex which targets HIF-1α for proteasomal degradation, consequently maintaining HIF-1α at low levels [51, 52]. Under hypoxic conditions, the prolyl hydroxylases are no longer active and the unmodified HIF-1α no longer interacts with VHL and thereby accumulates. The stabilized HIF-1α then translocates to the nucleus where it heterodimerizes with the HIF-1β subunit. The HIF-1 complex activates transcription by binding to a specific DNA sequence called Hypoxia Responsive Element (HRE) found within the promoter of target genes [53, 54].

Up to now, more than 70 genes are known to be transcriptionally activated by HIF-1 [55]. Several of these genes convey the survival and proliferation of tumor cells by mediating angiogenesis, glucose uptake, invasion and metastasis, thereby promoting tumorigenesis. Nuclear accumulation of HIF-1 protein has been reported in human common cancers and cells lines, including head and neck, glioblastomas, breast, colon, pancreatic and prostate, which are pathological conditions where furin was also found to be overexpressed [2, 5, 56–59]. In accordance with this, furin was recently identified as a novel HIF-1 target [19].

8.2   HIF-1 as a Regulator of the Furin Gene

Computer-assisted analysis of furin promoters P1, P1A and P1B uncovered the presence of several putative HREs similar to those found in many genes regulated by oxygen deprivation such as vascular endothelial growth factor, erythropoietin and glucose transporter-1 [53, 60, 61]. This suggested that furin expression may be regulated under hypoxic conditions to achieve adequate proteolytic maturation of angiogenic and/or tumorigenic substrates within tumors.

To test this hypothesis, HepG2 cells, derived from a human hepatoma, were cultured under hypoxic conditions (1% O_2). A marked increase in furin mRNA levels was observed with a maximum increase of 18-fold after 24 hours of low oxygen exposure. As a comparison, the convertases PACE4 and PC7 mRNA levels were barely affected by hypoxia, indicating that hypoxia-induced expression is not extended to all of the human proprotein convertase family members. Interestingly, a similar increase in furin mRNA was observed in other cell lines, including RAW 264.7 mouse macrophages, primary rat synoviocytes, as well as mouse Hepa-1 hepatoma cell line, indicating that the regulation of *fur* gene expression by hypoxia is extended to various cells types and species [19].

Several experiments were conducted in order to delineate the molecular mechanisms responsible for hypoxia-induced expression of furin [19]. The transient transfection of HepG2 cells with constructs encompassing the luciferase reporter gene under the control of each of the three furin promoters (P1, P1A and P1B) revealed that the *fur* P1 promoter is the most sensitive to hypoxia. Cotransfection of a HIF-1α dominant negative form in HepG2 cells or tranfection of the P1 promoter-reporter gene plasmid alone in HIF-1 deficient cells (Hepa-1c4 cells) indicated the requirement of HIF-1 for furin promoter activation by hypoxia. Further analysis of the hypoxia sensitive P1 promoter revealed that deletion of the –1221 to –413 region, containing two putative HREs, abolished the response to both hypoxia and to exogenous expression of HIF-1, suggesting that one or both of these HREs takes part in the furin transcriptional unit in response to hypoxia. Directed mutagenesis of each of these HREs permitted to define the critical HRE responsible for the HIF-1-dependent response of P1 promoter under hypoxia. This HRE, possessing enhancer capability, shares sequence similarities with those identified within the promoter of several tumor related genes, such as VEGF, MT1-MMP, and endoglin [44, 62, 63].

Several studies suggest that furin-induced expression within tumors supports tumoral development through the increased bioavailability of pro-tumorigenic mediators. Interestingly, the hypoxia/HIF-1-dependent increase in furin gene expression was found to be associated with increased maturation/activation of the angiogenic/tumorigenic mediators MT1-MMP and TGFβ1 [19]. Both mediators are well characterized furin substrates that have been found to profoundly affect many aspects of tumor progression [14, 17, 64]. MT1-MMP, through proteolytic events, regulates various cell functions, including extracellular matrix turnover, promotion of cell migration and invasion. MT1-MMP acts either through direct degradation of extracellular matrix components or indirectly by activating MMP-2 [65]. In addition, these metalloproteinases are involved in the angiogenic process by mediating endothelial cell invasion as well as the release or the activation of growth factors [66, 67]. TGFβ1, in turn, creates a favorable environment for tumor establishment by repressing immune surveillance, inducing the production of potent angiogenic factors such as vascular endothelial growth factor and basic fibroblast growth factor, and by increasing the production of extracellular matrix proteases, which promote tumor proliferation, invasion and metastasis [68, 69]. Moreover, in addition to these molecules, the multiplicity of other established furin substrates involved in cell growth and survival (insulin-like growth factor receptor-1, hepatocyte growth factor receptor c-Met, platelet-derived growth factor), cell invasion (E-cadherin and integrins), and angiogenesis (vascular endothelial growth factor-C) supports the contention that the regulation of furin activity within hypoxic/HIF-1 expressing zones of tumors could profoundly impact the course of tumor growth, invasion, and metastasis in a detrimental manner [70–75]. Thus, further studies are required in order to establish a link between increased maturation of furin substrates under hypoxic conditions and promotion of tumorigenesis. Preliminary experiments performed in our laboratory with the human fibrosarcoma cell line, HT1080, tend to confirm this link. Results obtained from fluorescent immuno-labelling of furin within mice xenografts, revealed a striking over-expression of furin within hypoxic areas of tumors, indicating that the proprotein convertase hypoxic regulation takes place *in vivo* (Grandmont S. et al., unpublished data).

Moreover, hypoxia was shown to be associated with an increase in cell invasive phenotype, in part through the up regulation of proteases, some of which are furin substrates [76, 77]. Interestingly, overexpression of the furin inhibitor α1-PDX in HT1080 cells blunted hypoxia-induced invasion to levels found in control cells

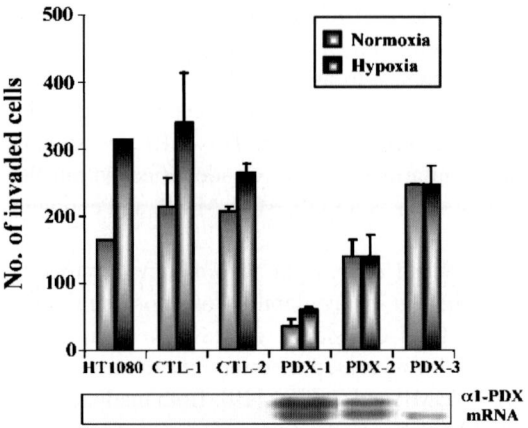

Figure 3. In vitro invasion was assessed in Boyden chambers using filters coated with type IV collagen, a major component of blood vessel basement membranes. Eighty thousand parental HT1080 cells or stable transfected cells with either the furin inhibitor α1-PDX (PDX-1, PDX-2, PDX-3) or the empty pcDNA3 vector (CTL-1 and CTL-2) were allowed to migrate under normoxic (21% O_2) or hypoxic (1% O_2) conditions for 8 hours. Cells that have reached the lower surface of filters were visually counted using a light microscope at 400-fold magnification. Northern blot analysis was performed to confirm the expression of furin inhibitor α1-PDX. Results are expressed as the mean ± SEM. A representative experiment out of three is shown

cultured in normoxia, suggesting that furin may significantly contribute to this process (Figure 3).

8.3   Furin Impacts Different Steps of HIF-1 Activity

Since the discovery of HIF-1, the scientific community has demonstrated a strong interest for the mechanisms that regulate the functions of this transcription factor. One of the most significant finding is the discovery that physiological stimuli other that hypoxia can regulate HIF-1 expression and activity. Several nonhypoxic stimuli, such as growth factors, hormones and cytokines were demonstrated to induce HIF-1α accumulation and activity in normal oxygen conditions, indicating that the tumor microenvironment encompasses, in addition to hypoxia, multiple checkpoints capable of regulating HIF-1 levels [78–82]. As opposed to hypoxia, stabilization does not seem to play a predominant role in the accumulation of HIF-1α in response to non-hypoxic stimuli. In fact, the HIF-1α degradation process is not altered in response to a variety of growth factors and cytokines [83–86]. The prime mechanisms underlying this induction involves the increase in HIF-1α gene transcription and/or mRNA translation, which seem sufficient to reverse the balance between synthesis and degradation, favoring HIF-1α accumulation in normoxia [83, 84, 87]. Although, not much is known about the mechanisms implicated in the transcriptional regulation of HIF-1α, several studies identified the PI3 kinase pathway and its downstream effectors as the mediators of increased HIF-1α translation in response to growth factors [88, 89].

8.4   Furin as a Regulator of HIF-1α Expression

Recent scientific developments indicate that furin can be regarded as a regulator of both HIF-1 expression and activity through its capacity to process/activate growth factors and receptors (Figure 4).

Among others, TGFβ1 was shown to induce the expression of HIF-1α under normoxic conditions leading to HIF-1 complex formation and induction of VEGF,

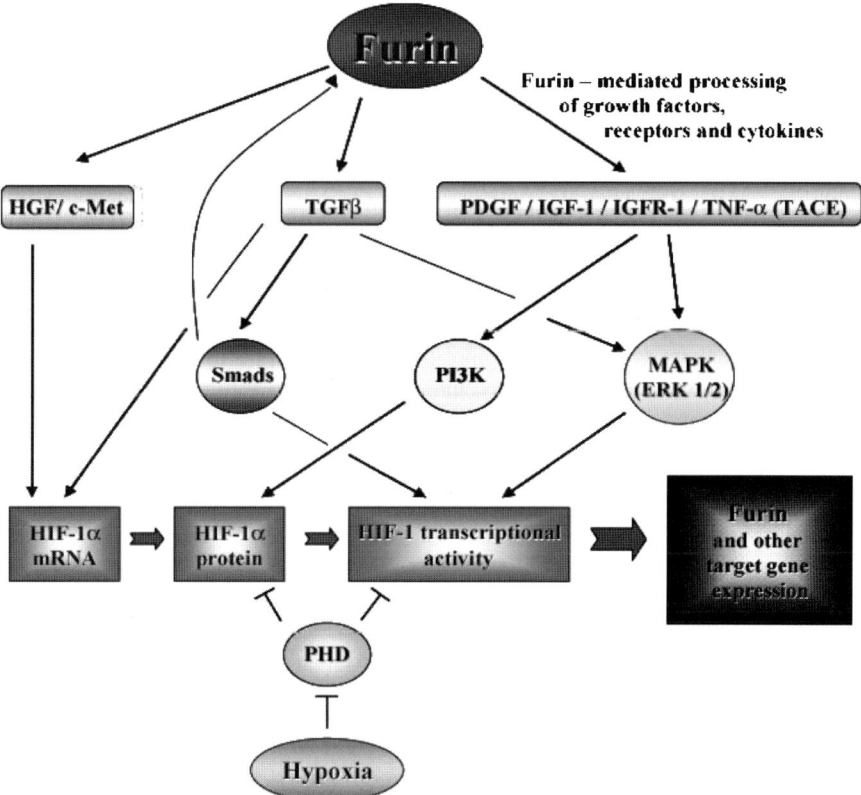

Figure 4. Schematic representation of furin impacts on HIF-1 expression and activity. Several furin substrates were shown to induce HIF-1α accumulation and activity in normoxic conditions. As opposed to hypoxia, that induces HIF-1α stabilization by repressing PHD activity, these non-hypoxic stimuli act by modifying HIF-1α transcription or translation levels through parallel or cooperative interactions between the Smads, PI3K and ERK 1/2 pathways. This will result in increased expression of the *fur* gene, a known HIF-1 target, creating a regulatory loop of potential importance in the induction as well as in the activation of numerous factors implicated in the pathology of cancer

a key target of HIF-1, indicating that the transcription factor is active [79, 90]. In
that study, however, the molecular mechanism involved in this increase of HIF-1α
expression was not explored. Since previous studies have indicated that increased
transcription plays a significant role in HIF-1α protein induction in response to
non-hypoxic stimuli, we evaluated whether a similar mechanism is involved in the
induction of HIF-1α by TGFβ1. Northern blot analysis revealed a marked induction
of HIF-1α mRNA levels in HepG2 cells after a 6h incubation in the presence of
TGFβ1, indicating that, as seen for other growth factors, TGFβ impacts HIF-1α
mRNA accumulation (McMahon et al., unpublished observation). Since this growth
factor is actively secreted by several cancer cells, furin expression in tumors could
be directly regulated by TGFβ1 through the action of Smads and/or HIF-1 [91].
This will result in an increased processing of the TGFβ1 precursor into its bioactive
form creating a regulatory cycle of potential importance in the induction, as well
as in the activation of numerous factors implicated in the pathology of cancer.

In fact, tumors express molecules, other than TGFβ1, that are known to induce
the expression of HIF-1α in normoxic conditions. Several are furin substrates such
as the growth factors IGF-1, HGF, and PDGF as well as the receptors IGFR1 and
c-Met, which transmit intracellular signals in response to IGF and HGF, respectively [71–73, 92, 93]. The inflammatory cytokines TNFα, which is liberated from
cell membranes through the action of another furin substrate, TACE, also induces
HIF-1 activity in normoxic conditions [93]. Thus, the activation of these molecules
by furin will likely confer to the convertase the ability to regulate HIF-1 expression,
which, in turn, could participate in the overexpression of furin, as well as other
HIF-1 targeted genes to exacerbate tumor progression. Supporting this, preliminary
results obtained from our laboratory indicate that furin inhibition is associated
with decreased basal levels of HIF-1 in cells, which is reverted by the addition of
exogenous growth factors (McMahon S. et al., unpublished data).

8.5   Furin as a Regulator of HIF-1 Activity

As mentioned, Smad proteins are known to regulate TGFβ-dependent gene
expression by interacting with a variety of co-activators, including FAST, AP-1,
SP1, as well as CBP/p300 [22, 32, 34, 35, 94, 95]. Several studies indicated that both
TGFβ1 and hypoxia signalling pathways can synergize, at the transcriptional level,
to induce the expression of erythropoietin, endoglin and VEGF genes [63, 96, 97].
This cooperative effect was attributed to the direct interaction between Smad3 and
HIF-1, as demonstrated by co-immunoprecipitation assays. Further experiments also
revealed the formation of a tripartite complex composed of SP1/HIF-1/Smad3 upon
cell exposition to both hypoxia and TGFβ [63]. Since TGFβ treatment or Smad3
transfection consistently resulted in increased HIF-1α/SP1 co-mmunoprecipitation,
Smad3 was regarded not only as a coactivator factor, but also as an adaptor molecule
that reinforce HIF-1/SP1 complex formation [63]. Because of its ability to bind to
general transcription factors as well as elements of the basal transcription machinery,
the presence of Sp1 in this multicomplex is though to be an additional way used by

HIF-1 to connect with theses molecules for more efficient transcriptional activity. In these studies, HIF-1/Smad cooperation was shown to be transmitted by the direct interaction of each transcription factor to respective and adjacent consensus sites [63, 96, 97]. Interestingly, the presence of putative Smad binding elements were uncovered surrounding the functional P1-HRE, indicating that in the context of the furin promoter, Smad may also serve as an adaptor to optimize furin expression.

In addition to hydroxylation, that controls HIF-1α stability, other post-translational modifications are necessary to confer maximal transcriptional activity to HIF-1. As an example, the c-terminal region of the HIF-1α subunit is directly phosphorylated by the MAPK p42/44 in vitro and in vivo, resulting in increased HIF-1 transcriptional activity [98, 99]. As opposed to hypoxia, several growth factors, including TGFβ1, are known to activate this signalling pathway following binding to their specific receptors [39]. Thus, furin can, through the bioactivation of TGFβ1, as well as other growth factors and receptors, enhance HIF-1 transcriptional activity in hypoxic and normoxic condition.

Together, these observations suggest that in addition to be regulated by HIF-1, furin can impact the expression and activity of this transcription factor. This sheds light on novel mechanisms by which the tumor microenvironment can modulate furin expression, but also, reveals a novel pro-tumorigenic action of furin, through the induction of HIF-1 expression and activity. This, in turn, can directly influence the expression of the convertase but also of several other HIF-1 regulated genes that strongly influence the tumorigenesis process.

9. EXPLOITING FURIN SIGNAL TRANSDUCTION PATHWAY FOR ANTICANCER THERAPY

In many cancer types, elevated furin expression is associated with the development of a more aggressive phenotype, a process that was found to be blocked by the expression of furin-specific inhibitors within tumor xenografts [5, 72, 100]. In homeostasis, however, low expression levels of furin are found in cells and tissues, where this convertase is involved in the maturation of proproteins needed for normal cellular functions. For this reason, long term uncontrolled inhibition of such an enzyme by sustained and unregulated expression of inhibitory proteins/drugs could likely present a significant risk of local effects. Recent findings demonstrating that hypoxic and non-hypoxic pathways are used by either tumor or inflammatory environments for the regulation of the furin convertase might uncover targeting therapeutic opportunities. For examples, hypoxic areas found in solid tumors as well as other common conditions such as arthritic joints and inflammatory bowel diseases can be useful for the specific targeting of anti-furin drugs via bioreductive prodrug delivery systems [101]. Similarly, the HIF/HRE transcription machinery is a powerful tool to target hypoxic and inflammatory areas of tumors with gene therapy. In several reports, the strong hypoxia-inducible enhancer of the erythropoietin gene was used with success to deliver the VEGF gene into ischemic myocardium as well as the suicide gene HSV-*tk* into human glioma and breast adenocarcinoma cells [102, 103]. Therefore, better understanding of molecules implicated in the signalling pathway leading to the expression of furin will become a priority for the development of strategies to block the *pathological facet* of this ubiquitously expressed convertase.

10. CONCLUSION

Recent advances in the signalling pathway that regulates furin expression have uncovered a complex interplay between Smad, ERK and HIF-1 pathways. How these components impact on furin expression during tumor development and how they interconnect with other signalling networks will still need to be addressed. Furthermore, cancer-type-specific differences in the activity of these signalling pathways might predict variations within the levels of furin expression among tumors. Therefore, there is no doubt that future studies will reveal further intricacies of signalling networks, with many more players and inter-relationships yet to be discovered.

ACKNOWLEDGEMENT

This work was supported by the Canadian Institute of Health Research.

REFERENCES

[1] Bassi DE, Fu J, Lopez de Cicco R, Klein-Szanto AJ (2005) Proprotein convertases: "master switches" in the regulation of tumor growth and progression. Mol Carcinog 44:151–161
[2] Cheng M, Watson PH, Paterson JA, Seidah N, Chretien M, Shiu RP (1997) Pro-protein convertase gene expression in human breast cancer. Int J Cancer 71:966–971
[3] Mbikay M, Sirois F, Yao J, Seidah NG, Chretien M (1997) Comparative analysis of expression of the proprotein convertases furin, PACE4, PC1 and PC2 in human lung tumours. Br J Cancer 75:1509–1514
[4] Bassi DE, Mahloogi H, Al-Saleem L, Lopez De Cicco R, Ridge JA, Klein-Szanto AJ (2001) Elevated furin expression in aggressive human head and neck tumors and tumor cell lines. Mol Carcinog 31:224–232
[5] Mercapide J, Lopez De Cicco R, Bassi DE, Castresana JS, Thomas G, Klein-Szanto AJ (2002) Inhibition of furin-mediated processing results in suppression of astrocytoma cell growth and invasiveness. Clin Cancer Res 8:1740–1746
[6] Khatib AM, Siegfried G, Chretien M, Metrakos P, Seidah NG (2002) Proprotein convertases in tumor progression and malignancy: Novel targets in cancer therapy. Am J Pathol 160:1921–1935
[7] Nakajima T, Konda Y, Kanai M, et al. (2002) Prohormone convertase furin has a role in gastric cancer cell proliferation with parathyroid hormone-related peptide in a reciprocal manner. Dig Dis Sci 47:2729–2737
[8] Wick W, Wild-Bode C, Frank B, Weller M (2004) BCL-2-induced glioma cell invasiveness depends on furin-like proteases. J Neurochem 91:1275–1283
[9] Elliott RL, Blobe GC (2005) Role of transforming growth factor Beta in human cancer. J Clin Oncol 23:2078–2093
[10] Roberts RB, Arteaga CL, Threadgill DW (2004) Modeling the cancer patient with genetically engineered mice: prediction of toxicity from molecule-targeted therapies. Cancer Cell 5:115–120

[11] Roberts AB, Wakefield LM (2003) The two faces of transforming growth factor beta in carcinogenesis. Proc Natl Acad Sci USA **100**:8621–8623

[12] Saharinen J, Hyytiainen M, Taipale J, Keski-Oja J (1999) Latent transforming growth factor-beta binding proteins (LTBPs)—structural extracellular matrix proteins for targeting TGF-beta action. Cytokine Growth Factor Rev **10**:99–117

[13] Annes JP, Munger JS, Rifkin DB (2003) Making sense of latent TGFbeta activation. J Cell Sci **116**:217–224

[14] Dubois CM, Blanchette F, Laprise MH, Leduc R, Grondin F, Seidah NG (2001) Evidence that furin is an authentic transforming growth factor-beta1-converting enzyme. Am J Pathol **158**:305–316

[15] Keski-Oja J, Koli K, von Melchner H (2004) TGF-beta activation by traction? Trends Cell Biol **14**:657–659

[16] Murphy-Ullrich JE, Poczatek M (2000) Activation of latent TGF-beta by thrombospondin-1: Mechanisms and physiology. Cytokine Growth Factor Rev **11**:59–69

[17] Dubois CM, Laprise MH, Blanchette F, Gentry LE, Leduc R (1995) Processing of transforming growth factor beta 1 precursor by human furin convertase. J Biol Chem **270**:10618–10624

[18] McMahon S, Laprise MH, Dubois CM (2003) Alternative pathway for the role of furin in tumor cell invasion process. Enhanced MMP-2 levels through bioactive TGFbeta. Exp Cell Res **291**:326–339

[19] McMahon S, Grondin F, McDonald PP, Richard DE, Dubois CM (2005) Hypoxia-enhanced expression of the proprotein convertase furin is mediated by hypoxia-inducible factor-1: Impact on the bioactivation of proproteins. J Biol Chem **280**:6561–6569

[20] Hua X, Liu X, Ansari DO, Lodish HF (1998) Synergistic cooperation of TFE3 and smad proteins in TGF-beta-induced transcription of the plasminogen activator inhibitor-1 gene. Genes Dev **12**:3084–3095

[21] Kim SJ, Angel P, Lafyatis R, et al. (1990) Autoinduction of transforming growth factor beta 1 is mediated by the AP-1 complex. Mol Cell Biol **10**:1492–1497

[22] Zhang Y, Feng XH, Derynck R (1998) Smad3 and Smad4 cooperate with c-Jun/c-Fos to mediate TGF-beta-induced transcription. Nature **394**:909–913

[23] Blanchette F, Rudd P, Grondin F, Attisano L, Dubois CM (2001) Involvement of Smads in TGFbeta1-induced furin (fur) transcription. J Cell Physiol **188**:264–273

[24] Blanchette F, Day R, Dong W, Laprise MH, Dubois CM (1997) TGFbeta1 regulates gene expression of its own converting enzyme furin. J Clin Invest **99**:1974–1983

[25] Attisano L, Wrana JL (1998) Mads and Smads in TGF beta signalling. Curr Opin Cell Biol **10**:188–194

[26] Massague J (1998) TGF-beta signal transduction. Annu Rev Biochem **67**:753–791

[27] Heldin CH, Miyazono K, ten Dijke P (1997) TGF-beta signalling from cell membrane to nucleus through SMAD proteins. Nature **390**:465–471

[28] Macias-Silva M, Abdollah S, Hoodless PA, Pirone R, Attisano L, Wrana JL (1996) MADR2 is a substrate of the TGFbeta receptor and its phosphorylation is required for nuclear accumulation and signaling. Cell **87**:1215–1224

[29] Souchelnytskyi S, Tamaki K, Engstrom U, Wernstedt C, ten Dijke P, Heldin CH (1997) Phosphorylation of Ser465 and Ser467 in the C terminus of Smad2 mediates interaction with Smad4 and is required for transforming growth factor-beta signaling. J Biol Chem **272**:28107–28115

[30] Zhang Y, Feng X, We R, Derynck R (1996) Receptor-associated Mad homologues synergize as effectors of the TGF-beta response. Nature **383**:168–172

[31] Nakao A, Imamura T, Souchelnytskyi S, et al. (1997) TGF-beta receptor-mediated signalling through Smad2, Smad3 and Smad4. Embo J **16**:5353–5362

[32] Kaestner KH, Knochel W, Martinez DE (2000) Unified nomenclature for the winged helix/forkhead transcription factors. Genes Dev **14**:142–146

[33] Chen X, Rubock MJ, Whitman M (1996) A transcriptional partner for MAD proteins in TGF-beta signalling. Nature **383**:691–696

[34] Attisano L, Wrana JL (2000) Smads as transcriptional co-modulators. Curr Opin Cell Biol **12**:235–243

[35] Liberati NT, Datto MB, Frederick JP, et al. (1999) Smads bind directly to the Jun family of AP-1 transcription factors. Proc Natl Acad Sci USA **96**:4844–4849

[36] Javelaud D, Mauviel A (2005) Crosstalk mechanisms between the mitogen-activated protein kinase pathways and Smad signaling downstream of TGF-beta: Implications for carcinogenesis. Oncogene **24**:5742–5750

[37] Ayoubi TA, Creemers JW, Roebroek AJ, Van de Ven WJ (1994) Expression of the dibasic proprotein processing enzyme furin is directed by multiple promoters. J Biol Chem **269**:9298–9303

[38] Dumaz N, Marais R (2005) Integrating signals between cAMP and the RAS/RAF/MEK/ERK signalling pathways. Based on the anniversary prize of the Gesellschaft fur Biochemie und Molekularbiologie Lecture delivered on 5 July 2003 at the Special FEBS Meeting in Brussels. Febs J **272**:3491–3504

[39] Blanchette F, Rivard N, Rudd P, Grondin F, Attisano L, Dubois CM (2001) Cross-talk between the p42/p44 MAP kinase and Smad pathways in transforming growth factor beta 1-induced furin gene transactivation. J Biol Chem **276**:33986–33994

[40] Liu J, Qu R, Ogura M, Shibata T, Harada H, Hiraoka M (2005) Real-time imaging of hypoxia-inducible factor-1 activity in tumor xenografts. J Radiat Res (Tokyo) **46**:93–102

[41] Vaupel P, Mayer A, Hockel M (2004) Tumor hypoxia and malignant progression. Methods Enzymol **381**:335–354

[42] Vaupel P, Mayer A (2005) Hypoxia and anemia: Effects on tumor biology and treatment resistance. Transfus Clin Biol **12**:5–10

[43] Rofstad EK, Danielsen T (1998) Hypoxia-induced angiogenesis and vascular endothelial growth factor secretion in human melanoma. Br J Cancer **77**:897–902

[44] Lal A, Peters H, St Croix B, et al. (2001) Transcriptional response to hypoxia in human tumors. J Natl Cancer Inst **93**:1337–1343

[45] Rofstad EK, Mathiesen B, Henriksen K, Kindem K, Galappathi K (2005) The tumor bed effect: Increased metastatic dissemination from hypoxia-induced up-regulation of metastasis-promoting gene products. Cancer Res **65**:2387–2396

[46] Kaur B, Khwaja FW, Severson EA, Matheny SL, Brat DJ, Van Meir EG (2005) Hypoxia and the hypoxia-inducible-factor pathway in glioma growth and angiogenesis. Neuro-oncol **7**:134–153

[47] Dachs GU, Patterson AV, Firth JD, et al. (1997) Targeting gene expression to hypoxic tumor cells. Nat Med **3**:515–520

[48] Hutchison GJ, Valentine HR, Loncaster JA, et al. (2004) Hypoxia-inducible factor 1alpha expression as an intrinsic marker of hypoxia: correlation with tumor oxygen, pimonidazole measurements, and outcome in locally advanced carcinoma of the cervix. Clin Cancer Res **10**:8405–8412

[49] Huang LE, Arany Z, Livingston DM, Bunn HF (1996) Activation of hypoxia-inducible transcription factor depends primarily upon redox-sensitive stabilization of its alpha subunit. J Biol Chem **271**:32253–32259

[50] Wang GL, Jiang BH, Rue EA, Semenza GL (1995) Hypoxia-inducible factor 1 is a basic-helix-loop-helix-PAS heterodimer regulated by cellular O2 tension. Proc Natl Acad Sci USA **92**:5510–5514

[51] Masson N, Willam C, Maxwell PH, Pugh CW, Ratcliffe PJ (2001) Independent function of two destruction domains in hypoxia-inducible factor-alpha chains activated by prolyl hydroxylation. Embo J **20**:5197–5206

[52] Epstein AC, Gleadle JM, McNeill LA, et al. (2001) C. elegans EGL-9 and mammalian homologs define a family of dioxygenases that regulate HIF by prolyl hydroxylation. Cell **107**:43–54

[53] Forsythe JA, Jiang BH, Iyer NV, et al. (1996) Activation of vascular endothelial growth factor gene transcription by hypoxia-inducible factor 1. Mol Cell Biol **16**:4604–4613

[54] Kimura H, Weisz A, Ogura T, et al. (2001) Identification of hypoxia-inducible factor 1 ancillary sequence and its function in vascular endothelial growth factor gene induction by hypoxia and nitric oxide. J Biol Chem **276**:2292–2298

[55] Semenza GL (2003) Targeting HIF-1 for cancer therapy. Nat Rev Cancer **3**:721–732

[56] Zhong H, De Marzo AM, Laughner E, et al. (1999) Overexpression of hypoxia-inducible factor 1alpha in common human cancers and their metastases. Cancer Res **59**:5830–5835

[57] Talks KL, Turley H, Gatter KC, et al. (2000) The expression and distribution of the hypoxia-inducible factors HIF-1alpha and HIF-2alpha in normal human tissues, cancers, and tumor-associated macrophages. Am J Pathol **157**:411–421

[58] Beasley NJ, Leek R, Alam M, et al. (2002) Hypoxia-inducible factors HIF-1alpha and HIF-2alpha in head and neck cancer: Relationship to tumor biology and treatment outcome in surgically resected patients. Cancer Res **62**:2493–2497

[59] Itoh Y, Tanaka S, Takekoshi S, Itoh J, Osamura RY (1996) Prohormone convertases (PC1/3 and PC2) in rat and human pancreas and islet cell tumors: Subcellular immunohistochemical analysis. Pathol Int **46**:726–737

[60] Chen C, Pore N, Behrooz A, Ismail-Beigi F, Maity A (2001) Regulation of glut1 mRNA by hypoxia-inducible factor-1. Interaction between H-ras and hypoxia. J Biol Chem **276**:9519–9525

[61] Semenza GL, Wang GL (1992) A nuclear factor induced by hypoxia via de novo protein synthesis binds to the human erythropoietin gene enhancer at a site required for transcriptional activation. Mol Cell Biol **12**:5447–5454

[62] Petrella BL, Lohi J, Brinckerhoff CE (2005) Identification of membrane type-1 matrix metalloproteinase as a target of hypoxia-inducible factor-2 alpha in von Hippel-Lindau renal cell carcinoma. Oncogene **24**:1043–1052

[63] Sanchez-Elsner T, Botella LM, Velasco B, Langa C, Bernabeu C (2002) Endoglin expression is regulated by transcriptional cooperation between the hypoxia and transforming growth factor-beta pathways. J Biol Chem **277**:43799–43808

[64] Yana I, Weiss SJ (2000) Regulation of membrane type-1 matrix metalloproteinase activation by proprotein convertases. Mol Biol Cell **11**:2387–2401

[65] Sounni NE, Janssen M, Foidart JM, Noel A (2003) Membrane type-1 matrix metalloproteinase and TIMP-2 in tumor angiogenesis. Matrix Biol **22**:55–61

[66] Sasaki K, Hattori T, Fujisawa T, Takahashi K, Inoue H, Takigawa M (1998) Nitric oxide mediates interleukin-1-induced gene expression of matrix metalloproteinases and basic fibroblast growth factor in cultured rabbit articular chondrocytes. J Biochem (Tokyo) **123**:431–439

[67] Jeong JW, Cha HJ, Yu DY, Seiki M, Kim KW (1999) Induction of membrane-type matrix metalloproteinase-1 stimulates angiogenic activities of bovine aortic endothelial cells. Angiogenesis **3**:167–174

[68] Pepper MS, Belin D, Montesano R, Orci L, Vassalli JD (1990) Transforming growth factor-beta 1 modulates basic fibroblast growth factor-induced proteolytic and angiogenic properties of endothelial cells in vitro. J Cell Biol **111**:743–755

[69] Sounni NE, Devy L, Hajitou A, et al. (2002) MT1-MMP expression promotes tumor growth and angiogenesis through an up-regulation of vascular endothelial growth factor expression. Faseb J **16**:555–564

[70] Siegfried G, Basak A, Cromlish JA, et al. (2003) The secretory proprotein convertases furin, PC5, and PC7 activate VEGF-C to induce tumorigenesis. J Clin Invest **111**:1723–1732

[71] Siegfried G, Basak A, Prichett Pejic W, et al. (2005) Regulation of the stepwise proteolytic cleavage and secretion of PDGF-B by the proprotein convertases. Oncogene **24**:6925–6935

[72] Khatib AM, Siegfried G, Prat A, et al. (2001) Inhibition of proprotein convertases is associated with loss of growth and tumorigenicity of HT-29 human colon carcinoma HT-29 cells: Importance of insulin-like growth factor-1 (IGF-1) receptor processing in IGF-1-mediated functions. J Biol Chem **276**:30686–30693

[73] Komada M, Hatsuzawa K, Shibamoto S, Ito F, Nakayama K, Kitamura N (1993) Proteolytic processing of the hepatocyte growth factor/scatter factor receptor by furin. FEBS Lett **328**:25–29

[74] Posthaus H, Dubois CM, Muller E (2003) Novel insights into cadherin processing by subtilisin-like convertases. FEBS Lett **536**:203–208

[75] Lissitzky JC, Luis J, Munzer JS, et al. (2000) Endoproteolytic processing of integrin pro-alpha subunits involves the redundant function of furin and proprotein convertase (PC) 5A, but not paired basic amino acid converting enzyme (PACE) 4, PC5B or PC7. Biochem J **346**:133–138

[76] Ridgway PF, Ziprin P, Alkhamesi N, Paraskeva PA, Peck DH, Darzi AW (2005) Hypoxia augments gelatinase activity in a variety of adenocarcinomas in vitro. J Surg Res **124**:180–186

[77] Graham CH, Forsdike J, Fitzgerald CJ, Macdonald-Goodfellow S (1999) Hypoxia-mediated stimu-
     lation of carcinoma cell invasiveness via upregulation of urokinase receptor expression. Int J
     Cancer **80**:617–623
[78] Richard DE, Berra E, Pouyssegur J (2000) Nonhypoxic pathway mediates the induction of
     hypoxia-inducible factor 1alpha in vascular smooth muscle cells. J Biol Chem **275**:26765–
     26771
[79] Gorlach A, Diebold I, Schini-Kerth VB, et al. (2001) Thrombin activates the hypoxia-inducible
     factor-1 signaling pathway in vascular smooth muscle cells: Role of the p22(phox)-containing
     NADPH oxidase. Circ Res **89**:47–54
[80] Jung Y, Isaacs JS, Lee S, Trepel J, Liu ZG, Neckers L (2003) Hypoxia-inducible factor induction
     by tumour necrosis factor in normoxic cells requires receptor-interacting protein-dependent nuclear
     factor kappa B activation. Biochem J **370**:1011–1017
[81] Haddad JJ, Land SC (2001) A non-hypoxic, ROS-sensitive pathway mediates TNF-alpha-dependent
     regulation of HIF-1alpha. FEBS Lett **505**:269–274
[82] Hellwig-Burgel T, Rutkowski K, Metzen E, Fandrey J, Jelkmann W (1999) Interleukin-1beta
     and tumor necrosis factor-alpha stimulate DNA binding of hypoxia-inducible factor-1. Blood
     94:1561–1567
[83] Page EL, Robitaille GA, Pouyssegur J, Richard DE (2002) Induction of hypoxia-inducible factor-
     1alpha by transcriptional and translational mechanisms. J Biol Chem **277**:48403–48409
[84] Tacchini L, De Ponti C, Matteucci E, Follis R, Desiderio MA (2004) Hepatocyte growth factor-
     activated NF-kappaB regulates HIF-1 activity and ODC expression, implicated in survival, differ-
     ently in different carcinoma cell lines. Carcinogenesis **25**:2089–2100
[85] Treins C, Giorgetti-Peraldi S, Murdaca J, Semenza GL, Van Obberghen E (2002) Insulin stimulates
     hypoxia-inducible factor 1 through a phosphatidylinositol 3-kinase/target of rapamycin-dependent
     signaling pathway. J Biol Chem **277**:27975–27981
[86] Fukuda R, Hirota K, Fan F, Jung YD, Ellis LM, Semenza GL (2002) Insulin-like growth factor 1
     induces hypoxia-inducible factor 1-mediated vascular endothelial growth factor expression, which
     is dependent on MAP kinase and phosphatidylinositol 3-kinase signaling in colon cancer cells.
     J Biol Chem **277**:38205–38211
[87] Blouin CC, Page EL, Soucy GM, Richard DE (2004) Hypoxic gene activation by lipopolysaccharide
     in macrophages: Implication of hypoxia-inducible factor 1alpha. Blood **103**:1124–1130
[88] Raught B, Gingras AC, Sonenberg N (2001) The target of rapamycin (TOR) proteins. Proc Natl
     Acad Sci USA **98**:7037–7044
[89] Gingras AC, Raught B, Sonenberg N (2001) Regulation of translation initiation by FRAP/mTOR.
     Genes Dev **15**:807–826
[90] Shih SC, Claffey KP (2001) Role of AP-1 and HIF-1 transcription factors in TGF-beta activation
     of VEGF expression. Growth Factors **19**:19–34
[91] Pasche B (2001) Role of transforming growth factor beta in cancer. J Cell Physiol **186**:153–168
[92] Duguay SJ, Lai-Zhang J, Steiner DF (1995) Mutational analysis of the insulin-like growth factor I
     prohormone processing site. J Biol Chem **270**:17566–17574
[93] Srour N, Lebel A, McMahon S, et al. (2003) TACE/ADAM-17 maturation and activation of
     sheddase activity require proprotein convertase activity. FEBS Lett **554**:275–283
[94] Poncelet AC, Schnaper HW (2001) Sp1 and Smad proteins cooperate to mediate transforming
     growth factor-beta 1-induced alpha 2(I) collagen expression in human glomerular mesangial cells.
     J Biol Chem **276**:6983–6992
[95] Pouponnot C, Jayaraman L, Massague J (1998) Physical and functional interaction of SMADs and
     p300/CBP. J Biol Chem **273**:22865–22868
[96] Sanchez-Elsner T, Botella LM, Velasco B, Corbi A, Attisano L, Bernabeu C (2001) Synergistic
     cooperation between hypoxia and transforming growth factor-beta pathways on human vascular
     endothelial growth factor gene expression. J Biol Chem **276**:38527–38535
[97] Sanchez-Elsner T, Ramirez JR, Sanz-Rodriguez F, Varela E, Bernabeu C, Botella LM (2004)
     A cross-talk between hypoxia and TGF-beta orchestrates erythropoietin gene regulation through
     SP1 and Smads. J Mol Biol **336**:9–24

[98] Richard DE, Berra E, Gothie E, Roux D, Pouyssegur J (1999) p42/p44 mitogen-activated protein kinases phosphorylate hypoxia-inducible factor 1alpha (HIF-1alpha) and enhance the transcriptional activity of HIF-1. J Biol Chem **274**:32631–32637

[99] Volmat V, Camps M, Arkinstall S, Pouyssegur J, Lenormand P (2001) The nucleus, a site for signal termination by sequestration and inactivation of p42/p44 MAP kinases. J Cell Sci **114**:3433–3443

[100] Lopez de Cicco R, Bassi DE, Zucker S, Seidah NG, Klein-Szanto AJ (2005) Human carcinoma cell growth and invasiveness is impaired by the propeptide of the ubiquitous proprotein convertase furin. Cancer Res **65**:4162–4171

[101] Seddon B, Kelland LR, Workman P (2004) Bioreductive prodrugs for cancer therapy. Methods Mol Med **90**:515–542

[102] Su H, Arakawa-Hoyt J, Kan YW (2002) Adeno-associated viral vector-mediated hypoxia response element-regulated gene expression in mouse ischemic heart model. Proc Natl Acad Sci USA **99**:9480–9485

[103] Greco O, Joiner MC, Doleh A, Powell AD, Hillman GG, Scott SD (2005) Hypoxia- and radiation-activated Cre/loxP 'molecular switch' vectors for gene therapy of cancer. Gene Ther **12**:974–979

CHAPTER 3

PACE4 GENE EXPRESSION IN HUMAN OVARIAN CANCER

BRIGITTE L. THÉRIAULT[1], YANGXIN FU[1], SHAWN K. MURRAY[2]
AND MARK W. NACHTIGAL[1,3]

[1] Department of Pharmacology
[2] Department of Pathology
[3] Department of Medicine, Tupper Medical Building, 5850 College Street, Dalhousie
University Halifax, NS, Canada, B3H 1X5

Abstract: Aberrant proprotein convertase (PC) activity is associated with human tumorigenesis. We recently determined that paired basic amino acid converting enzyme 4 (PACE4) expression is greatly reduced in metastatic human ovarian cancer cells compared to normal ovarian cells. Reduced PACE4 expression is due to epigenetic modification of the PACE4 promoter, and PACE4 expression can be stimulated in ovarian cancer cells by treatment with demethylating agents and histone deacetylase inhibitors. Stable PACE4 re-expression in ovarian cancer cells produces cellular senescence or reduced survival. Preliminary evidence from PACE4 mutant mice strongly suggest that loss of PACE4 production results in ovarian abnormalities, including reduced fertility, premature loss of follicle structures, and abnormal cell morphology with features reminiscent of ovarian tumors

Keywords: Ovarian cancer, PACE4, proprotein convertase, methylation, histone deacetylation

1.   PROPROTEIN CONVERTASES AND TUMOR CELL MALIGNANT PHENOTYPES

The proprotein convertase (PC) family of serine endoproteases plays a vital role in normal cellular physiology by converting proproteins to biologically active molecules. We have determined that one member of the human PC family, PACE4, is expressed in normal ovarian surface epithelial (OSE) cells, however this expression is greatly reduced in epithelial ovarian cancer (EOC) cells. We hypothesize that loss of PACE4 expression plays a role in development of human ovarian cancer.

A-Majid Khatib (ed.), Regulation of Carcinogenesis, Angiogenesis and Metastasis by the Proprotein Convertases, 47–65.

2.   BIOLOGY OF EPITHELIAL OVARIAN CANCER (EOC)

In this chapter we will restrict our discussion to EOC and will not discuss ovarian germ cell tumors or stromal tumors, which occur less frequently than EOC. EOC is comprised of five distinct histological subtypes (serous, clear cell, endometrioid, mucinous, and transitional) with serous adenocarcinomas occurring most frequently (Figure 1).

There is considerable debate regarding the origin of EOC, but it is believed that greater than 90% of ovarian tumors arise from the cells covering the surface of the ovary, the OSE (Figure 2; [1–4]). Our inabilities to conclusively ascribe the cellular origins of EOC stem from the fact that the etiology of ovarian cancer is poorly understood; there are currently no screening methods to detect ovarian cancer at an early stage of development.

Unlike many other human cancers, mouse models of human EOC have not been developed until recently [5–9]. A model of ovarian adenocarcinoma has been described in laying hens (Gallus domesticus) greater than 2 years of age [10], and spontaneous ovarian tumors can arise is some strains of mice [11] and rats [12]; however, these models show low tumor incidence and long latency periods, making them unattractive for experimental manipulation.

Evidence from human ovarian tissue samples suggests that OSE cells may become trapped within the ovarian cortex and form inclusion cysts (Figure 2C). Stratified cells that have lost their polarity and become dysplastic are often found in the inclusion cysts and may be a contributing factor in cancer formation [1, 13–15]. The OSE is usually separated from the stroma by a basement membrane comprised of laminin and collagen IV [16]. Epithelial cells on the ovarian surface are typically simple cuboidal (Figure 2A) or columnar (Figure 2B), whereas inclusion cysts are often lined with ciliated columnar epithelial cells that are not separated by a basement membrane and are in direct contact with stromal cells.

Current knowledge fails to explain the unique histological growth patterns of the different ovarian tumor subtypes. While direct experimental evidence defining factors that may contribute to OSE transformation are only now being identified, epidemiological evidence shows a correlation between increased parity and lactation, or oral contraceptive use for greater than 5 years, with a decreased risk of ovarian cancer. After ovulation the OSE divides and migrates to repair the wound left by the ovulated follicle. OSE cells may become encapsulated in the stroma and develop as inclusion cysts. Parity, lactation, and use of oral contraceptives coincide with reduced ovulation and a relatively quiescent OSE, thus reducing the chances of cyst formation. It remains unclear what factors influence the transformation of the OSE in the inclusion cysts, but one possibility is that these cells, once encased in the ovarian stroma, are subjected to a greater concentration of extracellular signaling molecules that may increase their chances for neoplastic transformation. While this has been a generally accepted theory for the cellular origin of ovarian cancer, this line of reasoning is compromised by the occurrence of ovarian cancer after prophylactic oophorectomy. Further research is required to ascertain if ovarian cancers may arise from cells such as fallopian tube or endometrial epithelium.

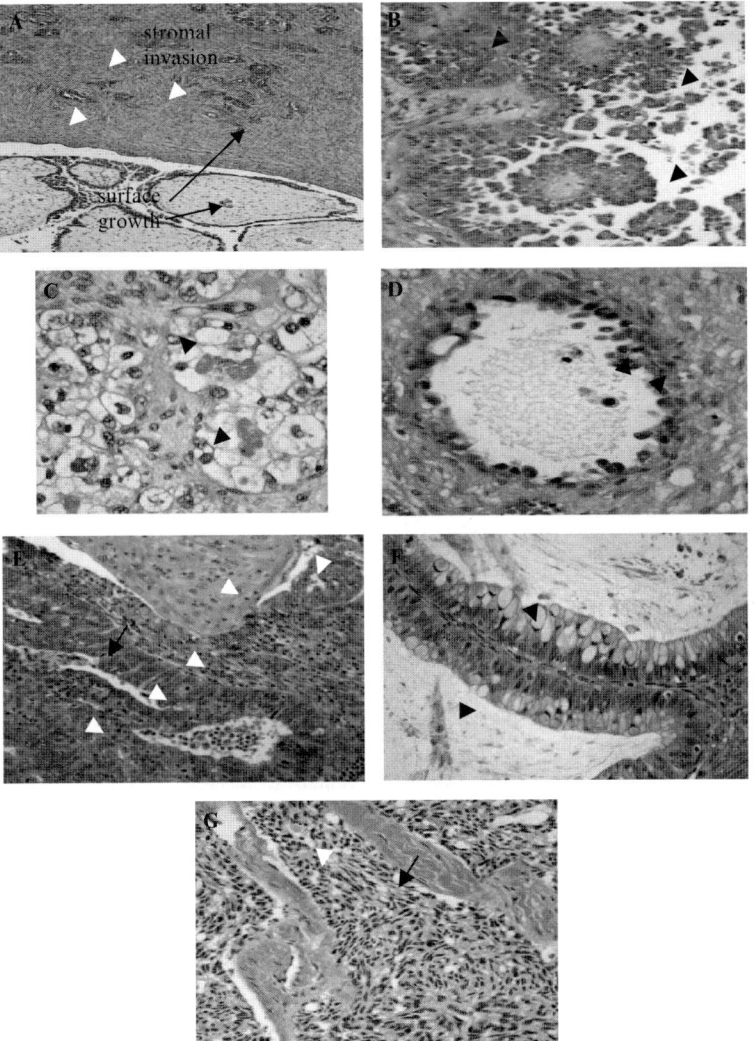

Figure 1. Epithelial ovarian cancer (EOC) is comprised of five distinct histological subtypes. (A) Ovarian serous adenocarcinoma. Note surface serous adenocarcinoma growth (arrows) and underlying invasive tumor infiltrating ovarian cortex (white arrowheads). This type of destructive stromal invasion is one feature that separates low-malignant potential or borderline tumors from frank carcinomas. (B) Hierarchical papillary clustering that is characteristic of ovarian serous carcinoma (black arrowheads). (C) Ovarian clear cell adenocarcinoma is characterized by solid collections of cells with clear cytoplasm and in the upper portion multiple rounded cytoplasmic concretions that may mimic alpha-fetoprotein globules of ovarian yolk sac tumor (black arrowheads). (D) Ovarian clear cell adenocarcinoma, note protrusion of nuclei into cystic space (so-called "hobnailing" of nuclei). (E) Ovarian endometrioid adenocarcinoma. Note endometrioid-type gland (white arrowheads) and nodules of squamous differentiation (arrow). (F) Ovarian mucinous adenocarcinoma. High power photomicrograph highlighting multiple neoplastic cells with ample cytoplasmic apical mucin vacuoles (black arrowheads). (G) Ovarian transitional cell carcinoma. Note large geographic sheets of slightly spindled and microcystic transitional or urothelial-type neoplastic epithelium (arrow) within a desmoplastic/fibrotic strom (white arrowhead)

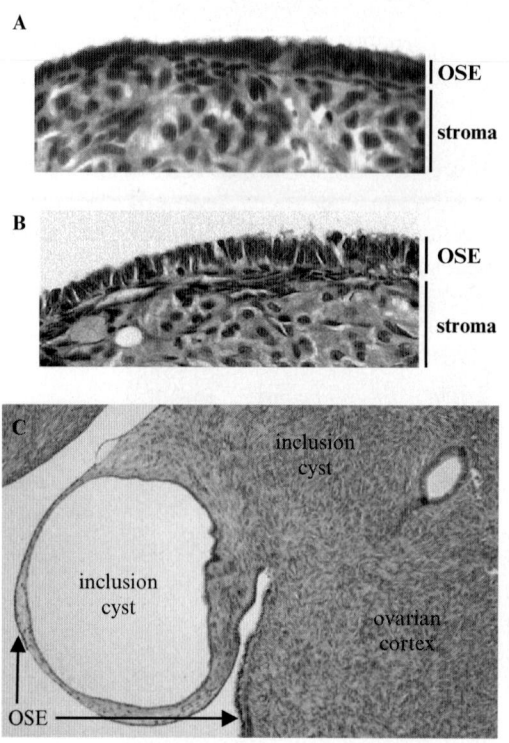

Figure 2. The ovarian surface epithelium (OSE). The surface of the ovary is covered by a single layer of epithelial cells termed the OSE. The OSE is separated from the underlying stromal cells and follicles by a basal lamina composed primarily of collagen type IV and laminin. OSE typically appears as cuboidal (A) or low columnar (B) epithelium. (C) Inclusion cysts lined with OSE cells within the ovarian cortex are the predominant sites of epithelial dysplasia

2.1   Modeling EOC

Non-familial (sporadic) disease accounts for ~90% of EOC, whereas the remaining 10% of EOC are familial [17, 18]. A variety of tumor suppressor genes associated with EOC have been identified by strategies such as positional cloning and differential expression techniques. Women carrying mutations in the breast cancer associated 1 (BRCA1) gene have an ~20–40% lifetime risk of developing EOC, whereas women with BRCA2 mutations harbor an ~10–20% risk [19]. BRCA carriers account for ~90% of familial ovarian cancers. *TP53* (p53) mutations are commonly found (>50%) in both familial and sporadic EOC [17, 18], and alterations in p53 occur frequently in the most common type of EOC, serous adenocarcinoma. Additionally, activating mutations, gene amplification, and loss of heterozygosity have been identified in sporadic EOC for *KRAS* [20], *AKT2* [21–23], *PTEN* [24, 25], and a number of other genes [26]. The study of human

EOC is challenging due to the fact that the field is just beginning to develop experimental animal models for epithelial ovarian tumorigenesis [5–9]. Tumor modeling by intraperitoneal (IP) injection of EOC cells is often used to replicate later stage ovarian cancer, which is characterized by tumor cell exfoliation and dissemination in the peritoneal cavity with formation of ascites fluid [27–29]. The lack of an OSE-specific promoter sequence to allow accurate targeting of transgenes to the OSE has been one of the factors contributing to a lag in our ability to model EOC. Recently the promoter for the Müllerian inhibiting substance (MIS) type II receptor (MISRII; a.k.a. Amhr2) was used to preferentially target the SV40 T antigen to mouse OSE, which resulted in the development of poorly differentiated epithelial ovarian carcinomas [6]. In addition to a TVA-replication-competent avian leucosis derived vector used for *ex vivo* manipulation of OSE [5], other methods have utilized viral technologies to specifically transduce the OSE after injection of the virus into the bursa surrounding the ovary [7, 9]. This technique allows highly specific infection of the OSE without affecting cells in the underlying ovarian stroma. Experimental evidence shows that cells in the bursa or oviductal epithelium were rarely transduced with the viral vector [9]. This method of viral transduction was used to generate OSE-specific knockouts by targeting deletion of floxed alleles after intrabursal injection of adenovirus expressing the Cre recombinase (Ad-CMV*Cre*) [7, 9]. The Cre recombinase efficiently produces recombination and deletion of floxed alleles after a single intrabursal injection of Ad-CMV*Cre*. The lab of Dr. T. Jacks produced the first rodent model of endometrioid ovarian cancer using intrabursal Ad CMV*Cre* injection to specifically activate K-ras^{G12D} and inactivate *Pten* in the mouse OSE [9]. A different model showed that deletion of floxed *p53* or *pRb* alone rarely produced ovarian tumours, but the combined deletion produced multiple serous cysts or poorly differentiated neoplasms in 100% of the animals (n = 34) [7]. In these ovarian cancer models mice displayed phenotypes ranging from undifferentiated carcinoma to well defined endometrioid adenocarcinoma, and the phenotype was dependent upon the combination of deleted tumour suppressor genes and activated oncogenes [7, 9]. Thus, the field is moving into an era of enhanced learning and experimental modeling that should contribute to advances in patient diagnoses and treatment.

3. PROPROTEIN CONVERTASES

The proprotein convertase (PC) family of calcium-dependent serine endoproteases plays a vital role in normal cellular physiology by converting proproteins to biologically active molecules. In mammals, nine PCs have been identified and cloned: furin, PC1/3, PC2, PC4, PACE4, PC5/6, PC7, SKI-1/SIP [30–33] and NARC-1 [34]. PCs are related to bacterial subtilisin, yeast kex2, and the *C. elegans* blisterase serine endoproteases that cleave precursor proteins at mono- and dibasic amino acid recognitions sites. The most recently cloned member, SKI-1/SIP, cleaves at non-basic residues. Paired basic amino acid converting enzyme 4 (PACE4) is a

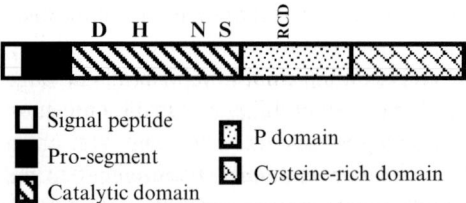

Signal peptide
Pro-segment
Catalytic domain
P domain
Cysteine-rich domain

Figure 3. Schematic diagram of PACE4 protein domains. Pace4 is a ~970 amino acid protein synthesized as a prepromolecule. The signal peptide and pro-segment are autoproteolytically cleaved in the
endoplasmic reticulum to produce mature Pace4. The catalytic domain utilizes a charge relay system
that is dependent on active site amino acid residues D, H, N, and S that are characteristic of serine
proteases. The P domain likely stabilizes the catalytic domain through hydrophobic interactions that
contribute to folding of the convertase; whereas the cysteine-rich domain is important for regulating
intracellular localization and secretion

~970 amino acid protein with a number of distinct domains critical for catalytic
activity, substrate interaction, and subcellular localization (Figure 3) [35].

PCs are packaged with their precursor substrates in the trans-Golgi network (TGN)
and cleavage can occur in the TGN or as the secretory vesicle is transported to the
plasma membrane. Furthermore, secreted forms of PCs can activate substrates in
the extracellular environment [36]. For example, Nodal expressed in the epiblast
and visceral endoderm is made bioactive by furin and Pace4 expressed by the
adjacent extraembryonic ectoderm [37]. *In vitro* PC molecules can display overlapping
substrate specificity; however, gene ablation studies clearly demonstrate that these
molecules are not 100% functionally redundant; furin [38] and Pace4 [39] null mice
share some degree of embryonic lethality, but display distinct phenotypes. Incom-
plete penetrance of the Pace4 null phenotype is thought to be due to reduced substrate
activity or partial compensation by furin *in vivo* [39]. Prior to 2003, there were
only two known cases of humans with a germ line PC defect, both inactivating
mutations in PC1/3 [40, 41]. Now, mutations in the gene encoding NARC-1 have
been found to contribute to autosomal dominant hypercholesterolemia; however,
the identification of NARC-1 substrates that contribute to the disease remain to
be elucidated [42, 43]. In addition to these patients, numerous reports demonstrate
that altered PC activity is associated with a variety of human tumours [44–49].

3.1   Is There a Role for PACE4 in Human Tumorigenesis?

We believe that de-regulated PACE4 activity is a critical factor contributing to
cellular alterations leading to a malignant tumor phenotype. It is unlikely that
PACE4 enzymatic activity would have a direct role in human tumor formation;
rather the altered substrate activity would be the causative agent in the tumor forming
process. Despite the fact that a wide variety of molecules that can participate in the
tumor forming process are targets for PC bioactivation (including growth factors
and their receptors, matrix metalloproteinases, and integrins [31], no *bona fide*

PACE4 substrates have been identified and shown to contribute to human tumorigenesis. Several reports show that PC overexpression correlates with more aggressive forms of human tumors, including head and neck squamous cell carcinoma [47], lung adenocarcinoma and squamous cell carcinoma [50], nonsmall cell lung carcinoma [44], and breast cancer [45]. In an animal model of chemically-induced squamous cell carcinoma conversion to spindle cell carcinoma, Hubbard et al. [46] found that PACE4 protein production was increased in the more malignant, poorly differentiated spindle cell carcinoma cells. Moreover, overexpression of PACE4 in squamous cell carcinoma cell lines with low invasive ability promoted their *in vivo* invasiveness. The most likely cause for increased invasiveness was through the observed gain of function to process prostromelysin-3 [51], a matrix metalloproteinase expressed in tissues undergoing active remodeling [52]. These studies highlight the need to develop substrate-screening assays that will help expand the field of PCs and cancer.

3.2   PC mRNA Expression in EOC Cells

As part of our ovarian cancer research program, we screened primary normal OSE and EOC cells from patients, and established EOC cell lines for PC expression. We found that these cells express PC5/6 (not shown), furin and PC7 (Figure 4A). As expected, none of these cells expressed PC1/3 or PC2, which are primarily produced by neuroendocrine cells (data not shown). The result that was most striking was the greatly reduced PACE4 mRNA expression in 9 primary EOC samples, 3 of 4 established EOC cell lines, and 2 immortalized 'normal' OSE cell lines (Figure 3A). This is in contrast to our data from 8 primary normal OSE patient samples that expressed PACE4 mRNA. Interestingly, this was one of the first reports indicating reduced PACE4 expression in a human cancer [53]. We confirmed this result using quantitative RT-PCR analysis (Figure 4B).

Our results are in agreement with previously published microarray data examining gene expression in EOC [54]. PACE4 was identified as a gene preferentially expressed in normal OSE compared with EOC-derived epithelial cells, with ~0.2 fold PACE4 expression in EOC cells (*p*-value, 0.03). In addition, there were no significant changes in furin (1.0 fold, *p*-value, 0.8), PC5/6 (1.2 fold, *p*-value, 0.18), or PC7 (1.46 fold, *p*-value, 0.09) mRNA expression (D. Matei, personal communication). The majority of reports to date have demonstrated that overexpression of PCs contribute to adverse tumor cell biology, including enhancing migration, invasion, and metastasis. We believe that reduced PC activity is equally likely to produce a disruption in the regulation of cellular activities that may promote tumorigenesis. Evidence of this behavior was demonstrated recently by Nejjari et al. [55] when they inhibited PC activity in HT29-D4 colon adenocarcinoma cells using either α1-antitrypsin Portland (a furin/PC6B/PACE4 selective inhibitor) or the furin inhibitor decRVKR-CMK. They found that cells where PC activity was reduced resulted in enhanced cell migration, *in vitro* invasion, and *in vivo* metastases formation. Addition of these inhibitors blocked endoproteolytic cleavage of αv

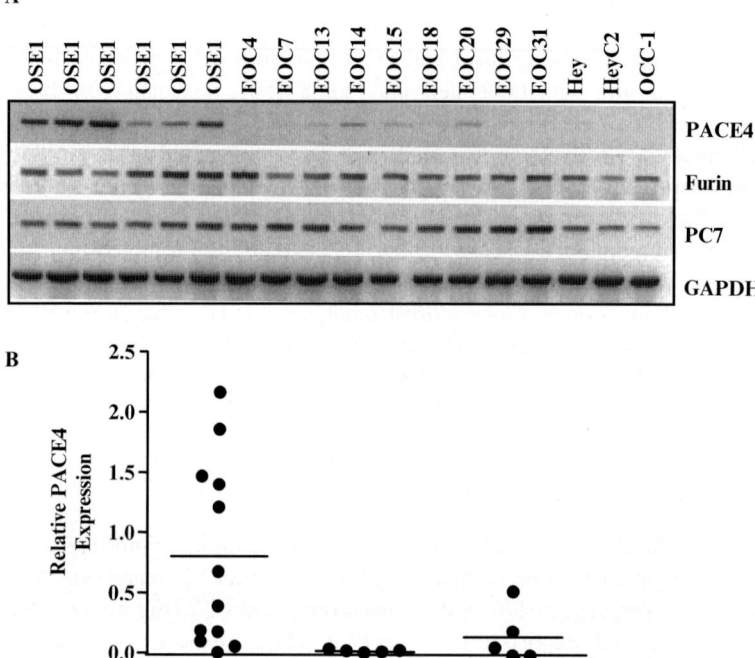

Figure 4. PACE4 expression is reduced in EOC cells compared to normal OSE. (A) Proprotein convertase PACE4, Furin, and PC7 mRNA expression was assessed in normal OSE (OSE1-6), primary EOC cells (EOC#), and EOC cell lines (Hey, HeyC2, OCC-1) by RT-PCR. GAPDH was used as a PCR gel loading control. [Reprinted with permission from *Molecular Cancer Research*]. (B) Quantitative RT-PCR was used to measure relative levels of PACE4 in normal OSE, primary EOC cells, and EOC cell lines (SkOV3, HeyC2, CaOV3, OvCa429, OCC-1). GAPDH was used as an internal control for data normalization. The horizontal bar indicates the mean level of expression, with the range of expression for normal OSE = 0.004 to 2.16, primary EOC = 0.013 to 0.041, and EOC cell lines = 0.001 to 0.534. Amongst the EOC cell lines, PACE4 expression was highest in OvCa429 (0.534) and CaOV3 (0.201)

integrin, which is critical for adhesion through interaction with extracellular matrix proteins. Nejjari et al., also treated IGROV1 ovarian cancer cells with decRVKR-CMK, which resulted in enhanced cell migration, presumably due to the effects on reduced cell adhesion [55]. Collectively, these data suggest that fine control of PC protein levels is required to maintain cellular homeostasis, and that disruption of PC levels contribute to human tumour formation.

3.3   The PACE4 Promoter is Active in EOC Cells

Southern analysis of genomic DNA from EOC cells allowed us to eliminate the possibility that gross PACE4 gene (*PCSK6*) rearrangement or loss accounted for reduced PACE4 gene expression (data not shown). Therefore we analyzed whether

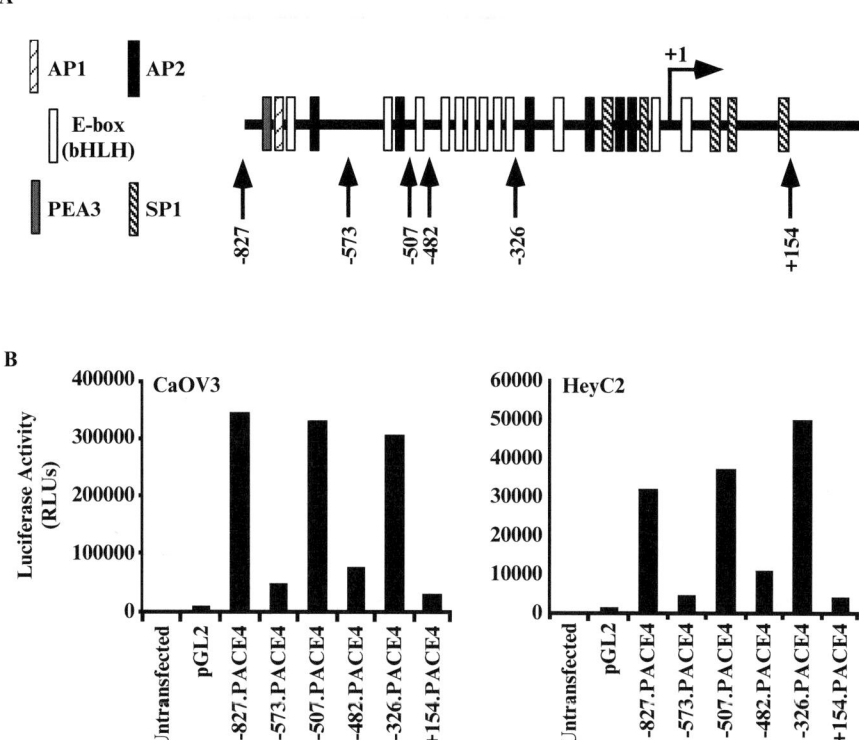

Figure 5. Human PACE4 promoter is active in EOC cell lines. (A) The human PACE4 5′-flanking DNA and first exon have many putative transcription factor binding sites that include AP1, AP2, PEA3, SP1, and E-box binding sites for basis helix-loop-helix (bHLH) factors [modified from Tsuji et al. [38]]. +1 is defined as the start of transcription (indicated by horizontal arrow). The human PACE4 promoter and first exon (−827 to +314) was PCR amplified and subcloned into the pGL2.basic luciferase reporter plasmid. Promoter deletion constructs used in (*B*) are indicated by vertical arrows. (B) Representative transfection data. Ovarian cancer cell lines were transfected with PACE4 reporter gene constructs. Luciferase activity is reported as relative light units (RLUs). pGL2.basic is included as a negative control; data was normalized to β-gal activity produced by co trasfection with the internal reference plasmid pCMV.β-galactosidase

reduced PACE4 mRNA expression in EOC cells was due to altered PACE4 promoter activity. The human PACE4 promoter has been cloned [56] and can be regulated in different cell types by *achaete-scute* homolog 1 and 2 basic helix-loop-helix transcription factors binding to a number of E-box elements [57–59]. The PACE4 promoter and part of exon 1 (−837 to +315) was PCR amplified and cloned into a luciferase reporter plasmid (Figure 5A).

The full length PACE4 promoter construct produced high levels of luciferase activity in CaOV3, HeyC2 and OCC-1 EOC cell lines, indicating that the PACE4 promoter can be activated in these cells (Figure. 5B) [53]. Moreover, deletion analysis identified regions of 5′-flanking DNA required to maintain high levels of promoter activity, similar to the results of Tsuji et al. [57]. Unlike Tsuji et al. [57], where PACE4 promoter activity was examined in rat pituitary GH_4C1 and human liver HepG2 cells, we found that deletion of nucleotides −827 to −573 produced a decrease in PACE4 promoter activity, while deletion of nucleotides −573 to −507 restored activity. These results suggest that a tissue-specific repressor element may be localized between nucleotides −573 and −507. Overall, our results show that the PACE4 promoter is active in EOC cells, making it unlikely that reduced PACE4 expression is due to a loss of transcription factors necessary for PACE4 expression.

3.4   Methylation Status of PACE4 Promoter and Exon I in Normal OSE and EOC Cells

The observation that established EOC cell lines have reduced PACE4 expression, but maintained the ability to support PACE4 promoter activity, suggested that reduced expression may be due to epigenetic modification of the PACE4 gene, such as promoter DNA methylation or histone deacetylation. The DNA sequence of the PACE4 promoter region and first exon is highly GC rich [56], and contains a CpG island that provides a putative target for aberrant alteration of chromatin architecture leading to gene silencing. The methylation status of the endogenous PACE4 promoter and exon I CpG rich region (−196 to +340) was assessed by bisulfite genomic sequencing [53]. Genomic DNA from two normal OSE, the established HeyC2 cell line, and four primary EOC samples was modified by bisulfite treatment. Eight clones from each cell sample were selected for sequence analysis and 79 CpG dinucleotides in the PACE4 gene (−196 to −25 in 5′-flanking DNA; +9 to +340 in exon I) were examined. Normal OSE had 8–9% methylation in the 79 CpG dinucleotides examined, whereas the methylated CpG cytosine nucleotides in EOC cells ranged from 58–93% (Table 1). These data show that the PACE4 gene is hypermethylated in primary EOC cells and established cell lines compared with normal OSE, suggesting that DNA hypermethylation plays a role in the reduction of PACE4 expression in EOC cells.

3.5   PACE4 Gene Expression is Greatly Increased by Demethylation or Inhibition of Histone Deacetylation

Epigenetic alterations of chromatin structure, including DNA hypermethylation and histone deacetylation, can contribute to gene silencing. To study whether epigenetic alterations of the PACE4 gene contribute to the reduced PACE4 mRNA expression, we initially treated HeyC2 cells with the demethylating agent 5-aza-dC (Aza), the histone deacetylase inhibitor trichostatin A (TSA), or both [53]. After treatment, the level of PACE4 expression was quantified by Southern analysis of RT-PCR products amplified using primers specific to the PACE4 cDNA (nts. 2896–3308). If promoter/exon I hypermethylation and histone deacetylation play a role in reduction of PACE4 expression, treatment of these cells with these inhibitors will increase or

Table 1. Methylation status of endogenous PACE4 promoter and exon 1 (nucleotides −196 to +340). Percent of methylated cytosine in the 70 CpG dinucleotides is shown. [Data reproduced with permission from *Molecular Cancer Research*]

Sample	% methylation
OSE6	9.2
OSE7	8.2
HeyC2	92.6
EOC4	92.9
EOC7	58.1
EOC13	74.4
EOC15	62.0

restore PACE4 mRNA expression. As shown in Figure 6, both Aza (1 or 10 μM) and TSA (0.1 or 1 μM) treatment resulted in increased PACE4 expression. In addition, a synergistic effect between Aza and TSA was observed in HeyC2 cells at all doses (Figure 6A).

The effect of each drug alone was maximal at the highest concentration tested, therefore 10 μM Aza and 1 μM TSA were used to study the effect of these drugs on PACE4 expression in OCC-1 cells and 3 primary EOC cell samples. To assess whether the treatments were specific for PACE4, PC7 expression was also analyzed. Aza treatment alone increased PACE4 expression in all cells (Figure 6B). Although TSA alone failed to increase PACE4 expression in one of these cell samples (EOC14), a synergistic effect between Aza and TSA was observed in all cells; for example PACE4 expression was increased 22.6 fold in OCC-1 cells. Minimal increases were observed for PC7 expression, with a 1.6 fold increase as the greatest response. These data indicate that promoter hypermethylation and histone deacetylation can play a role in the reduced expression of PACE4 in EOC cells.

DNA hypermethylation and histone deacetylation are frequent epigenetic events in cancer that are associated with gene silencing [60, 61]. GC-rich sequences, also known as CpG islands, are often found in promoter DNA and are targets of DNA methylation. Methyl CpG binding proteins (MBD1-1 and MeCP2) [62] can recruit histone deacetylases that act locally to produce a condensed chromatin structure that leads to transcriptional repression [63]. In normal cells CpG methylation plays an important role in regulating gene expression, whereas in cancer cells aberrant promoter methylation or hypermethylation can lead to abnormal gene silencing, including repression of tumor suppressor genes. Studies to assess the patterns of CpG island methylation in human EOC (over 100 samples) using either methylation specific PCR [64, 65] or differential methylation hybridization [66, 67] revealed frequently methylated CpG islands in many loci associated with human tumor formation, including the *BRCA1* gene, the putative tumor suppressor gene *HIC1*, and *MLH1*. Moreover, expression of DNA methyltransferases (DNMT1 and DNMT3b) is elevated in established EOC cell lines compared to normal OSE [68]. These data strongly suggest that DNA methylation contributes to gene silencing in EOC cells. Our experiments

A

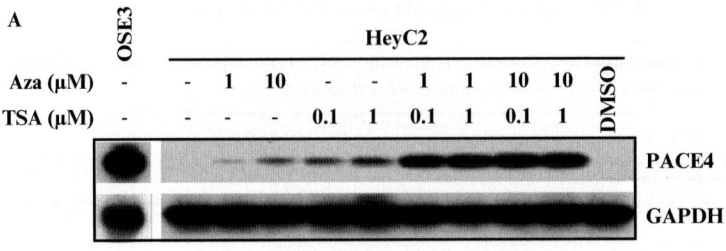

B

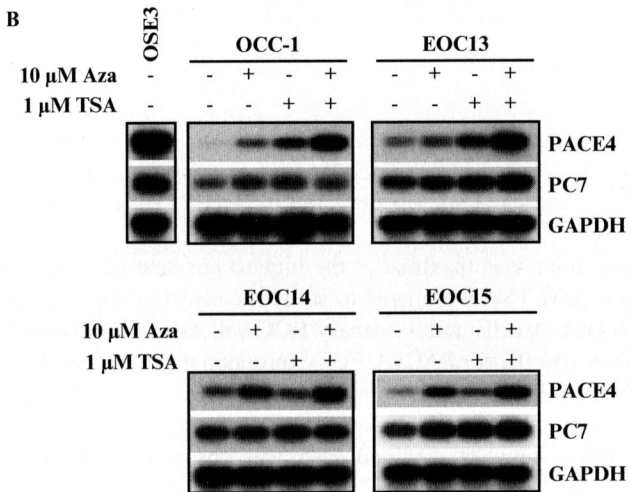

Figure 6. PACE4 gene expression is increased by treatment with a demethylating agent and/or a histone deacetylase inhibitor. (A) HeyC2 cells were treated with Aza ($1\,\mu$M or $10\,\mu$M) or TSA ($0.1\,\mu$M or $1\,\mu$M) or a combination of various doses of Aza and TSA. Cells were cultured in the medium containing Aza for 3 days, with medium changes every day. TSA was applied for the last 18 h. PACE4 expression was examined using Southern analysis of RT-PCR products (isolated after 20 PCR cycles) using the full-length PACE4 cDNA as a probe. DMSO was used as a vehicle control. Expression in untreated normal OSE (OSE3) was used as a positive control. (B) OCC-1 and three primary EOC cell samples (EOC13-15) were untreated (−) or treated (+) with $10\,\mu$M Aza, $1\,\mu$M TSA or both, and PACE4, PC7, or GAPDH expression was examined. Expression in untreated normal OSE (OSE3) was used as a positive control. [Reprinted with permission from *Molecular Cancer Research*]

support the notion that both PACE4 (*PCSK6*) DNA hypermethylation and histone deacetylation contribute to the reduction of PACE4 gene expression in EOC cells.

3.6   PACE4 Re-expression in EOC Cells

We followed up these observations by examining the consequence of PACE4 re-expression in EOC cells. Interestingly, we generated stable EOC cell lines expressing wild type PACE4 resulting in numerous recombinant lines that expressed

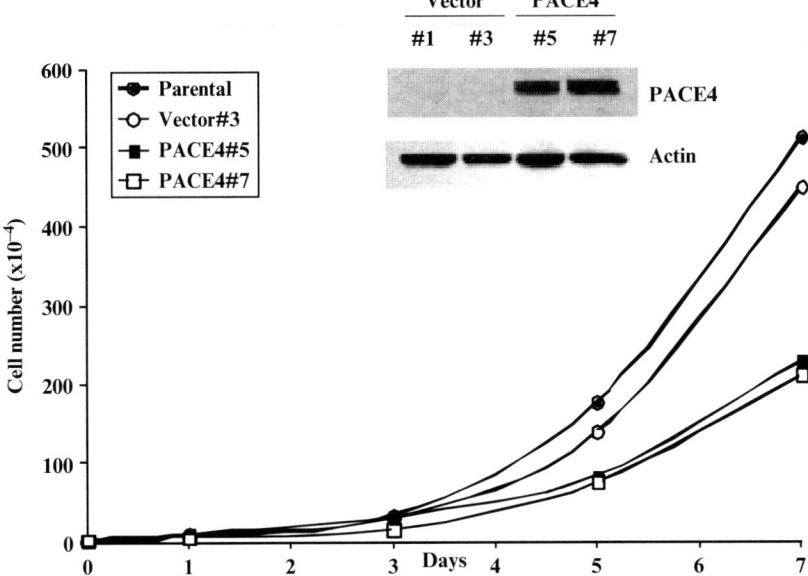

Figure 7. PACE4 re-expression in EOC cell lines reduces cell proliferation. HeyC2 cells were transfected with pcDNA3.1-myc-his (Vector; Invitrogen) or pcDNA3.1-PACE4-myc-his (PACE4) and selected with G418. G418 resistant clones were amplified and PACE4 expression assessed using anti-myc antibody by Western analysis (inset; actin was used as a loading control). An equal number of cells were plated on day 0, and grown for 7 days. Cells were counted every two days. Compared to parental or vector control cells, there are fewer HeyC2 cells in clones re-expressing PACE4

PACE4 and produced a dramatic decrease in EOC proliferation (Figure 7). However, all of these lines eventually became senescent or died after short-term mainte-nance in culture. Similar results were obtained with clones from HeyC2, OCC-1, OvCa429, and CaOV3 EOC cell lines. These results strongly suggest that EOC cells are very sensitive to bioactivation of PACE4 substrates. Our experiments to generate stable PACE4 expressing clones utilized constitutive (pCMV) promoter systems, therefore, future experiments should explore the utilization of inducible expression systems, or clones containing flox-STOP sequences removable with Cre recombinase, to evaluate the role that PACE4 may play in modulating the tumorigenic behavior of EOC cells.

4.   CAN WE LEARN SOMETHING ABOUT OVARIAN CANCER BIOLOGY FROM PACE4 MUTANT MICE?

Constam and Robertson generated Pace4 mutant (Pace4$^{-/-}$) mice to investigate the role of the BMP/TGFβ superfamily of growth factors in mammalian organo-genesis [39]. Morphogenic gradients of BMP-like molecules (nodal, lefty, and BMP4) establish antero-posterior and left-right axis formation in vertebrates.

Genetic analysis indicates that furin and Pace4 are upstream of nodal and BMP4 [37, 39, 69]. In Pace4$^{-/-}$ mice, loss of substrate bioactivation can result in embryonic lethality around embryonic day (e) 13.5 due to severe cardiac malformations [39]. We recently obtained Pace4$^{-/-}$ mice to further our analysis of the contribution of PACE4 biology to ovarian tumorigenesis. Similar to the original report [39] we have observed that the average number of Pace4$^{-/-}$ pups found dead at birth was ∼22% (male and female; our data obtained from >350 births), compared to 2.5% (n = 155) and 9% (n = 293) neonatal deaths in wild type and heterozygous animals, respectively. There is no obvious morphological defect in Pace4$^{-/-}$ mice younger than 6 months of age. These mice are fertile and produce the expected number of pups per birth when the ∼25% embryonic lethality is considered (average litter size is 6.3, n = 53 litters). We did observe that the number of days between births was increased in many Pace4$^{-/-}$ animals compared to wild type animals (∼37 days for Pace4$^{-/-}$ versus ∼22 days for wild type), suggesting that there may be a defect in reproductive endocrine physiology in Pace4$^{-/-}$ mice. However, our preliminary studies show dramatic differences between older Pace4$^{-/-}$ mice (7.5–9 months of age) and age-matched wild type (+/+) or heterozygous (+/−) animals upon gross anatomical and morphological evaluation of the ovaries. While some Pace4$^{-/-}$ mice (3 of 8) exhibited relatively normal ovaries (similar size, follicles of different developmental stages, numerous corpora lutea), the majority of the Pace4$^{-/-}$ mice (5 of 8) had ovaries that were notably different. The most prominent abnormalities were diminished or absent follicular development and increased inclusions of OSE; features consistent with premature ovarian failure and precancerous ovarian lesions. The most severely affected ovaries exhibited a complete lack of granulosa cell aggregates, no discernable nests, cords or collections of granulosa cells, and consequently no organized follicles or evidence of germ cells. This was accompanied by accentuated invaginations or inclusions of OSE/mesothelium into the ovarian parenchyma producing a complex tubular configuration. Other areas exhibited loss of polarity within the cell lining with diminished apical cytoplasm and protruding nuclei giving a somewhat "hobnailed" appearance. A tubal phenotype and loss of cell polarity including loss of columnar cell shape and hobnailed nuclei are prominent features of serous carcinoma and clear cell carcinoma as opposed to the well-oriented columnar cells in endometrioid and mucinous carcinomas of the endometrium or ovary, tumors with a much better prognosis. OSE alterations in the Pace4$^{-/-}$ mice indicate that Pace4 substrates may play a role in cell growth and orientation within the OSE. This is exciting as this model may help us understand some of the mechanisms of OSE alteration involved in tumor evolution and the resultant histologic subtype (e.g., serous carcinoma). Consequently, we believe that the Pace4$^{-/-}$ mice constitute an important tool for the study of ovarian cancer.

Changes were also present in the uterine lining of Pace4$^{-/-}$ mice. Noted was a diminished proliferative response in the endometrium with reduced endometrial glandular elements, more prominent stroma, and endometrial surface epithelium lacking stratification or mitotic activity, in contrast to controls. The uterine appearance is consistent with hypoestrogenemia, likely caused by reduced follicle number. Failure of ovarian function and infertility lead to atrophy of the endometrium and secondary Müllerian system [70]; an association of increased risk for EOC occurs with a history of infertility and some conditions including endometriosis [71]. There is significant evidence that the majority of cases of ovarian endometriosis likely represent a metaplastic alteration of the OSE [72] and ovarian endometriosis may undergo a variety of cytologic alterations [73–75], including tubal/ciliated alteration and 'hobnail' alteration, similar to those identified in the Pace4$^{-/-}$ mice. Future work will attempt to elucidate mechanisms by which there is an increased risk of ovarian and endometrial cancer with infertility and aging. We are continuing to evaluate ovarian and uterine morphology to develop a more complete picture regarding the nature of the alterations and determine the timing of the changes in cellular morphology in the Pace4$^{-/-}$ animals.

5.   WHERE DO WE GO FROM HERE?

The study of PCs and cancer is an emerging field. In many respects it is surprising that this subject has remained relatively understudied, given the obvious role that PCs play in fundamental cell metabolism and regulation. Perhaps one of the difficulties contributing to the expansion of this field has been the lack of large-scale substrate screening methods. Historically, investigators have used dedicated recombinant PC cleavage assays to determine which PCs were capable of cleaving known putative target proproteins, and subsequently defining these targets as PC substrates. Developing high-throughput screening technologies to identify PC substrates are critical to advance this field and provide insight into the molecules that contribute to normal and pathologic cellular processes. These technologies could include proteomic approaches involving substrate co-purification (co-immunoprecipitation, GST-pull down, or TAP-tag technology) coupled with mass spectrometry to identify the substrates, or perhaps non-traditional approaches including genetic screens using yeast, bacterial, or mammalian two-hybrid methods. Identifying PACE4 substrates will give us insight into the possible role that these molecules play in the pathobiology associated with altered PACE4 expression. The focus of our research efforts will be to understand how reduced PACE4 contributes to the development of ovarian cancer, and may lead to the identification of pathways involved in the formation of different ovarian carcinoma subtypes.

ACKNOWLEDGEMENTS

The authors wish to acknowledge Drs. R. Grimshaw and J. Bentley (Gynecologic Oncology, QEII Health Science Centre) for providing human ovarian tumor samples, Dr. T. F. Baskett and the staff of the QEII Ob/Gyn Unit for providing normal human ovarian tissue, and Dr. D. Matei for the microarray data. This work was supported by a Regional Partnership grant to M.W.N from the Canadian Institutes of Health Research (MT-15438), the Nova Scotia Health Research Foundation (NSHRF) and the Dalhousie Cancer Research Program. M.W.N. is a Research

Scientist of the Canadian Cancer Society through an award from the National Cancer Institute of Canada, and B.L.T is supported by a Studentship from the NSHRF.

REFERENCES

[1] Scully RE (1995) Pathology of ovarian cancer precursors. Journal of Cellular Biochemistry 23:208–218

[2] Dubeau L (1999) The cell of origin of ovarian epithelial tumors and the ovarian surface epithelium dogma: Does the emperor have no clothes? Gynecol Oncol 72:437–442

[3] Auersperg N, Wong AS, Choi KC, Kang SK, Leung PC (2001) Ovarian surface epithelium: Biology, endocrinology, and pathology. Endocrine Reviews 22:255–288

[4] Horiuchi AK, Itoh M, Shimizu I, Nakai T, Yamazaki K, Kimura A, Suzuki I, Shiozawa N Ueda, Konishi I (2003) Toward understanding the natural history of ovarian carcinoma development: A clinicopathological approach. Gynecol Oncol 88:309–317

[5] Orsulic S, Li Y, Soslow RA, Vitale-Cross LA, Gutkind JS, Varmus HE (2002) Induction of ovarian cancer by defined multiple genetic changes in a mouse model system. Cancer Cell 1:53–62

[6] Connolly DC, Bao R, Nikitin AY, Stephens KC, Poole TW, Hua X, Harris SS, Vanderhyden BC, Hamilton TC (2003) Female mice chimeric for expression of the simian virus 40 TAg under control of the MISIIR promoter develop epithelial ovarian cancer. Cancer Res 63:1389–1397

[7] Flesken-Nikitin A, Choi KC, Eng JP, Shmidt EN, Nikitin AY (2003) Induction of carcinogenesis by concurrent inactivation of p53 and Rb1 in the mouse ovarian surface epithelium. Cancer Res 63:3459–3463

[8] Liu J, Yang G, Thompson-Lanza JA, Glassman A, Hayes K, Patterson A, Marquez RT, Auersperg N, Yu Y, Hahn WC, Mills GB, Bast RC Jr (2004) A genetically defined model for human ovarian cancer. Cancer Res 64:1655–1663

[9] Dinulescu DM, Ince TA, Quade BJ, Shafer SA, Crowley D, Jacks T (2005) Role of K-ras and Pten in the development of mouse models of endometriosis and endometrioid ovarian cancer. Nat Med 11:63–70

[10] Fredrickson TN (1987) Ovarian tumors of the hen. Environ Health Perspect. 73:35–51

[11] Tillmann T, Kamino K, Mohr U (2000) Incidence and spectrum of spontaneous neoplasms in male and female CBA/J mice. Exp Toxicol Pathol 52:221–225

[12] Gregson RL, Lewis DJ, Abbott DP (1984) Spontaneous ovarian neoplasms of the laboratory rat. Vet Pathol. 21:292–299

[13] Scully RE, Young RH, Clement PB (1996) Tumors of the ovary, maldeveloped gonads, fallopian tube, and broad ligament. In: Rosai, J (ed), Atlas of Tumor Biology, Armed Forces Institute of Pathology, Washington, D.C.

[14] Radisavljevic S (1976) The pathogenesis of ovarian inclusion cysts and cystomas. Obstet Gynecol 49:424–429

[15] Hamilton T, Becker JL, Kacinski BM, Nicosia SV, Rodriguez G (2003) Discussion: Development of experimental models for ovarian cancer. Gynecol Oncol 88:S52–S55

[16] Yang DH, Smith ER, Cohen C, Wu H, Patriotis C, Godwin AK, Hamilton TC, Xu XX (2002) Molecular events associated with dysplastic morphologic transformation and initiation of ovarian tumorigenicity. Cancer 94:2380–2392

[17] Lynch HT, Casey MJ, Lynch J, White TEK, Godwin AK (1998) Genetics and ovarian cancer. Seminars in Oncology 25:265–280

[18] Russell SEH (2000) The Molecular Pathogenesis of Ovarian Cancer. In: Ovarian Cancer: Methods and Protocols, Bartlett J.M.S., (ed) Humana Press, Inc., Totowa, NJ, pp 25–36

[19] Werness BA, Eltabbakh GH (2001) Familial ovarian cancer and early ovarian cancer: Biologic, pathologic, and clinical features. Int J Gynecol Pathol 20:48–63

[20] Mok SC, Bell DA, Knapp RC, Fishbaugh PM, Welch WR, Muto MG, Berkowitz RS, Tsao SW (1993) Mutation of K-ras protooncogene in human ovarian epithelial tumors of borderline malignancy. Cancer Res 53:1489–1492

[21] Cheng JQ, Godwin AK, Bellacosa A, Taguchi T, Franke TF, Hamilton TC, Tsichlis PN, Testa JR (1992) AKT2, a putative oncogene encoding a member of a subfamily of protein-serine/threonine kinases, is amplified in human ovarian carcinomas. Proc Natl Acad Sci U S A **89**:9267–9271

[22] Bellacosa A, de Feo D, Godwin AK, Bell DW, Cheng JQ, Altomare DA, Wan M, Dubeau L, Scambia G, Masciullo V, et al. (1995) Molecular alterations of the AKT2 oncogene in ovarian and breast carcinomas. Int J Cancer **64**:280–285

[23] Liu AX,Testa JR, Hamilton TC, Jove R, Nicosia SV, Cheng JQ (1998) AKT2, a member of the protein kinase B family, is activated by growth factors, v-Ha-ras, and v-src through phosphatidyli-nositol 3-kinase in human ovarian epithelial cancer cells. Cancer Res **58**:2973–2977

[24] Saito M, Okamoto A, Kohno T, Takakura S, Shinozaki H, Isonishi S, Yasuhara T, Yoshimura T, Ohtake Y, Ochiai K, Yokota J, Tanaka T (2000) Allelic imbalance and mutations of the PTEN gene in ovarian cancer. Int J Cancer **85**:160–165

[25] Kurose K, Zhou XP, Araki T, Cannistra SA, Maher ER, Eng C (2001) Frequent loss of PTEN expression is linked to elevated phosphorylated Akt levels, but not associated with p27 and cyclin D1 expression, in primary epithelial ovarian carcinomas. Am J Pathol **158**:2097–2106

[26] Gray JW, Suzuki S, Kuo WL, Polikoff D, Deavers M, Smith-McCune K, Berchuk A, Pinkel D, Albertson D, Mills GB (2003) Specific Keynote: Genome copy number abnormalities in ovarian cancer. Gynecol Oncol **88**:16–21

[27] Yu D, Wolf JK, Scanlon M, Price JE, Hung MC (1993) Enhanced c-erbB-2/neu expression in human ovarian cancer cells correlates with more severe malignancy that can be suppressed by E1A. Cancer Res **53**:891–898

[28] Ong A, Maines-Bandiera SL, Roskelley CD, Auersperg N (2000) An ovarian adenocarcinoma line derived from SV40/E-cadherin-transfected normal human ovarian surface epithelium. International Journal of Cancer **85**:430–437

[29] Selvakumaran M, Pisarcik DA, Bao R, Yeung AT, Hamilton TC (2003) Enhanced cisplatin cytotoxicity by disturbing the nucleotide excision repair pathway in ovarian cancer cell lines. Cancer Res. **63**:1311–1316

[30] Seidah NG, Chretien M (1999) Proprotein and prohormone convertases: A family of subtilases generating diverse bioactive polypeptides. Brain Res. **848**:45–62

[31] Khatib AM, Siegfried G, Chretien M, Metrakos P, Seidah NG (2002) Proprotein convertases in tumor progression and malignancy: Novel targets in cancer therapy. Am J Pathol **160**:1921–1935

[32] Czyzyk TA, Morgan DJ, Peng B, Zhang J, Karantzas A, Arai M, Pintar JE (2003) Targeted mutagenesis of processing enzymes and regulators: Implications for development and physiology. J Neurosci Res **74**:446–455

[33] Taylor NA, Van De Ven WJ, Creemers JW (2003) Curbing activation: Proprotein convertases in homeostasis and pathology. Faseb J **17**:1215–1227

[34] Seidah NG, Benjannet S, Wickham L, Marcinkiewicz J, Jasmin SB, Stifani S, Basak A, Prat A, Chretien M (2003) The secretory proprotein convertase neural apoptosis-regulated convertase 1 (NARC-1): Liver regeneration and neuronal differentiation. Proc Natl Acad Sci U S A **100**:928 933

[35] Kiefer MC, Tucker JE, Joh R, Landsberg KE, Saltman D, Barr PJ (1991) Identification of a second human subtilisin-like protease gene in the fes/fps region of chromosome 15. DNA Cell Biol **10**:757–769

[36] Tsuji A, Sakurai K, Kiyokage E, Yamazaki T, Koide S, Toida K, Ishimura K, Matsuda Y (2003) Secretory proprotein convertases PACE4 and PC6A are heparin-binding proteins which are localized in the extracellular matrix. Potential role of PACE4 in the activation of proproteins in the extracellular matrix. Biochim Biophys Acta **1645**:95–104

[37] Beck S, Le Good JA, Guzman M, Ben Haim N, Roy K, Beermann F, Constam DB (2002) Extraembryonic proteases regulate Nodal signalling during gastrulation. Nat Cell Biol **4**:981–985

[38] Roebroek AJ, Umans L, Pauli IG, Robertson EJ, van Leuven F, Van de Ven WJ, Constam DB (1998) Failure of ventral closure and axial rotation in embryos lacking the proprotein convertase Furin. Development **125**:4863–4876

[39] Constam DB, Robertson EJ (2000) SPC4/PACE4 regulates a TGFbeta signaling network during axis formation. Genes Dev **14**:1146–1155

[40] O'Rahilly S, Gray H, Humphreys PJ, Krook A, Polonsky KS, White A, Gibson S, Taylor K, Carr C (1995) Brief report: Impaired processing of prohormones associated with abnormalities of glucose homeostasis and adrenal function. N Engl J Med 333:1386–1390

[41] Jackson RS, Creemers JW, Ohagi S, Raffin-Sanson ML, Sanders L, Montague CT, Hutton JC, O'Rahilly S (1997) Obesity and impaired prohormone processing associated with mutations in the human prohormone convertase 1 gene. Nat Genet 16:303–306

[42] Abifadel M, Varret M, Rabes JP, Allard D, Ouguerram K, Devillers M, Cruaud C, Benjannet S, Wickham L, Erlich D, Derre A, Villeger L, Farnier M, Beucler I, Bruckert E, Chambaz J, Chanu B, Lecerf JM, Luc G, Moulin P, Weissenbach J, Prat A, Krempf M, Junien C, Seidah NG, Boileau C (2003) Mutations in PCSK9 cause autosomal dominant hypercholesterolemia. Nat Gene 34:154–156

[43] Benjannet S, Rhainds D, Essalmani R, Mayne J, Wickham L, Jin W, Asselin MC, Hamelin J, Varret M, Allard D, Trillard M, Abifadel M, Tebon A, Attie AD, Rader DJ, Boileau C, Brissette L, Chretien M, Prat A, Seidah NG (2004) NARC-1/PCSK9 and its natural mutants: Zymogen cleavage and effects on the low density lipoprotein (LDL) receptor and LDL cholesterol. J Biol Chem. 279:48865–48875

[44] Schalken JA, Roebroek AJ, Oomen PP, Wagenaar SS, Debruyne FM, Bloemers HP, Van de Ven WJ (1987) Fur gene expression as a discriminating marker for small cell and nonsmall cell lung carcinomas. J Clin Invest 80:1545–1549

[45] Cheng M, Watson PH, Paterson JA, Seidah N, Chretien M, Shiu RP (1997) Pro-protein convertase gene expression in human breast cancer. Int J Cancer 71:966–971

[46] Hubbard FC, Goodrow TL, Liu SC, Brilliant MH, Basset P, Mains RE, Klein-Szanto AJ (1997) Expression of PACE4 in chemically induced carcinomas is associated with spindle cell tumor conversion and increased invasive ability. Cancer Res 57:5226–5231

[47] Bassi DE, Lopez De Cicco R, Mahloogi H, Zucker S, Thomas G, Klein-Szanto AJ (2001) Furin inhibition results in absent or decreased invasiveness and tumorigenicity of human cancer cells. Proc Natl Acad Sci U S A 98:10326–10331

[48] Khatib AM, Siegfried G, Prat A, Luis J, Chretien M, Metrakos P, Seidah NG (2001) Inhibition of proprotein convertases is associated with loss of growth and tumorigenicity of HT-29 human colon carcinoma cells: Importance of insulin-like growth factor-1 (IGF-1) receptor processing in IGF-1-mediated functions. J Biol Chem 276:30686–30693

[49] Siegfried G, Basak A, Cromlish JA, Benjannet S, Marcinkiewicz J, Chretien M, Seidah NG, Khatib AM (2003) The secretory proprotein convertases furin, PC5, and PC7 activate VEGF-C to induce tumorigenesis. J Clin Invest 111:1723–1732

[50] Mbikay M, Sirois F, Yao J, Seidah NG, Chretien M (1997) Comparative analysis of expression of the proprotein convertases furin, PACE4, PC1 and PC2 in human lung tumours. Br J Cancer 75:1509–1514

[51] Mahloogi H, Bassi DE, Klein-Szanto AJ (2002) Malignant conversion of non-tumorigenic murine skin keratinocytes overexpressing PACE4. Carcinogenesis 23:565–572

[52] Pei D, Weiss SJ (1995) Furin-dependent intracellular activation of the human stromelysin-3 zymogen. Nature 375:244–247

[53] Fu Y, Campbell EJ, Shepherd TG, Nachtigal MW (2003) Epigenetic regulation of proprotein convertase PACE4 gene expression in human ovarian cancer cells. Mol Cancer Res 1:569–576

[54] Matei D, Graeber TG, Baldwin RL, Karlan BY, Rao J, Chang DD (2002) Gene expression in epithelial ovarian carcinoma. Oncogene 21:6289–6298

[55] Nejjari M, Berthet V, Rigot V, Laforest S, Jacquier MF, Seidah NG, Remy L, Bruyneel E, Scoazec JY, Marvaldi J, Luis J (2004) Inhibition of proprotein convertases enhances cell migration and metastases development of human colon carcinoma HT-29 cells in a rat model. Am J Pathol 164:1925–1933

[56] Tsuji A, Hine C, Tamai Y, Yonemoto K, Mori K, Yoshida S, Bando M, Sakai E, Akamatsu T, Matsuda Y (1997) Genomic organization and alternative splicing of human PACE4 (SPC4), kexin-like processing endoprotease. J Biochem (Tokyo) 122:438–452

[57] Tsuji A, Yoshida S, Hasegawa S, Bando M, Yoshida I, Koide S, Mori K, Matsuda Y (1999) Human subtilisin-like proprotein convertase, PACE4 (SPC4) gene expression is highly regulated through E-box elements in HepG2 and GH4C1 cells. J Biochem (Tokyo) **126**:494–502

[58] Yoshida I, Koide S, Hasegawa SI, Nakagawara A, Tsuji A, Matsuda Y (2001) Proprotein convertase PACE4 is down-regulated by the basic helix-loop-helix transcription factor hASH-1 and MASH-1. Biochem J **360**:683–689

[59] Koide S, Yoshida I, Tsuji A, Matsuda Y (2003) The expression of proprotein convertase PACE4 is highly regulated by Hash-2 in placenta: Possible role of placenta-specific basic helix-loop-helix transcription factor, human achaete-scute homologue-2. J Biochem **13**:433–440

[60] Herman JG, Baylin SB (2003) Gene silencing in cancer in association with promoter hypermethylation. N Engl J Med **349**:2042–2054

[61] Egger G, Liang G, Aparicio A, Jones PA (2004) Epigenetics in human disease and prospects for epigenetic therapy. Nature **429**:457–463

[62] Bird AP, Wolffe AP (1999) Methylation-induced repression—belts, braces, and chromatin. Cell **99**:451–454

[63] Leonhardt H, Cardoso MC (2000) DNA methylation, nuclear structure, gene expression and cancer. J Cell Biochem **35**:78–83

[64] Herman JG, Graff JR, Myohanen S, Nelkin BD, Baylin SB (1996) Methylation-specific PCR: A novel PCR assay for methylation status of CpG islands. Proc Natl Acad Sci U S A **93**:9821–9826

[65] Strathdee G, Appleton K, Illand M, Millan DW, Sargent J, Paul J, Brown R (2001) Primary ovarian carcinomas display multiple methylator phenotypes involving known tumor suppressor genes. Am J Pathol. **158**:1121–1127

[66] Richon VM, Sandhoff TW, Rifkind RA, Marks P.A (2000) Histone deacetylase inhibitor selectively induces p21WAF1 expression and gene-associated histone acetylation. Proc Natl Acad Sci U S A **97**:10014–10019

[67] Osada H, Tatematsu Y, Masuda A, Saito T, Sugiyama M, Yanagisawa K, Takahashi T (2001) Heterogeneous transforming growth factor (TGF)-beta unresponsiveness and loss of TGF-beta receptor type II expression caused by histone deacetylation in lung cancer cell lines. Cancer Res **61**:8331–8339

[68] Ahluwalia AP, Yan JA, Hurteau RM, Bigsby SH, Jung TH, Huang, Nephew KP (2001) DNA methylation and ovarian cancer. I. Analysis of CpG island hypermethylation in human ovarian cancer using differential methylation hybridization. Gynecol Oncol. **82**:261–268

[69] Constam DB, Robertson EJ (1999) Regulation of bone morphogenetic protein activity by pro domains and proprotein convertases. J Cell Biol **144**:139–149

[70] Lauchlan SC (1972) The secondary Mullerian system. Obstet Gynecol Surv **27**:133–146

[71] Brinton LA, Lamb EJ, Moghissi KS, Scoccia B, Althuis MD, Mabie JE, and Westhoff CL (2004) Ovarian cancer risk associated with varying causes of infertility. Fertil Steril. **82**:405–414

[72] El-Mahgoub S, Yaseen S (1980) A positive proof for the theory of coelomic metaplasia. Am J Obstet Gynecol. **137**:137–140

[73] Lauchlan SC (1966) The cytology of endometriosis. Am J Obstet Gynecol **94**:533–535

[74] Czernobilsky B, Morris WJ (1979) A histologic study of ovarian endometriosis with emphasis on hyperplastic and atypical changes. Obstet Gynecol **53**:318–323

[75] Fukunaga M, Ushigome S (1998) Epithelial metaplastic changes in ovarian endometriosis. Mod Pathol **11**:784–788

CHAPTER 4

PROPROTEIN CONVERTASES IN TUMORIGENESIS, ANGIOGENESIS AND METASTASIS

GERALDINE SIEGFRIED[1], MICHEL CHRÉTIEN[2] AND
ABDEL-MAJID KHATIB[3]

[1] *INSERM U143, Le Kremlin-Bicêtre, France*
[2] *Regional Protein Chemistry Centre, Diseases of Ageing Unit, Ottawa Health Research Institute, Loeb Building, 725 Parkdale Ave., Ottawa, Ontario, Canada*
[3] *INSERM U 716/AVENIR. Institut de Génétique Moléculaire, Laboratoire de Pharmacologie Expérimentale et Clinique Paris, France*

Abstract: Proprotein convertases (PCs) are directly responsible for the activation of protein precursors implicated in neoplasia by either the degradation of extracellular matrix and modulation of cell adhesion, growth and/or survival. These functions are crucial in the acquisition of the tumorigenic phenotype, tumor progression and metastasis. Here we discuss a number of recent findings on the role of these enzymes in the regulation of multiple cellular functions that impact on the invasive/metastatic potential of cancer cells

Keywords: Proliferation, adhesion, survival, tumorigenesis, metastasis, angiogenesis

1. PROPROTEIN CONVERTASES AND TUMOR CELL MALIGNANT PHENOTYPES

Tumor progression is characterized by the gradual acquisition of various cellular aberrations resulting in malignant behavior such as lack of control in cell proliferation and survival. Similarly, cancer metastasis is a complex process involving sequential interactions between the disseminating tumor cells and a continuously changing host microenvironment. The formation of metastases usually resulted from the detachment of a cell or group of tumor cells from the primary tumor that migrate and invade the local host tissue. The newly formed lesion can itself become the source of disseminating cells and give rise to secondary and tertiary metastasis (Figure 1).

A-Majid Khatib (ed.), Regulation of Carcinogenesis, Angiogenesis and Metastasis by the Proprotein Convertases, 67–88.
© 2006 *Springer.*

These processes are mediated by molecular interactions resulting from the dissolution of tissue barriers through the degradative activity of enzymes such as proteinases, growth modulation by paracrine, host-derived peptide growth factors, and cell-cell or cell-extracellular matrix communication through specific adhesion receptors, such as integrins, which regulate tumor cell attachment, spreading, and migration [1–3]. Recently many of the molecules that regulate these processes revealed to be activated and secreted via series of endoproteolytic steps involving, among others, the proprotein convertases (PCs) [4], thereby involving directly the activity of these enzymes in the control of the malignant phenotypes and metastatic potential of tumor cells, such cellular growth, survival, invasiveness and tumorigenesis (Figure 1).

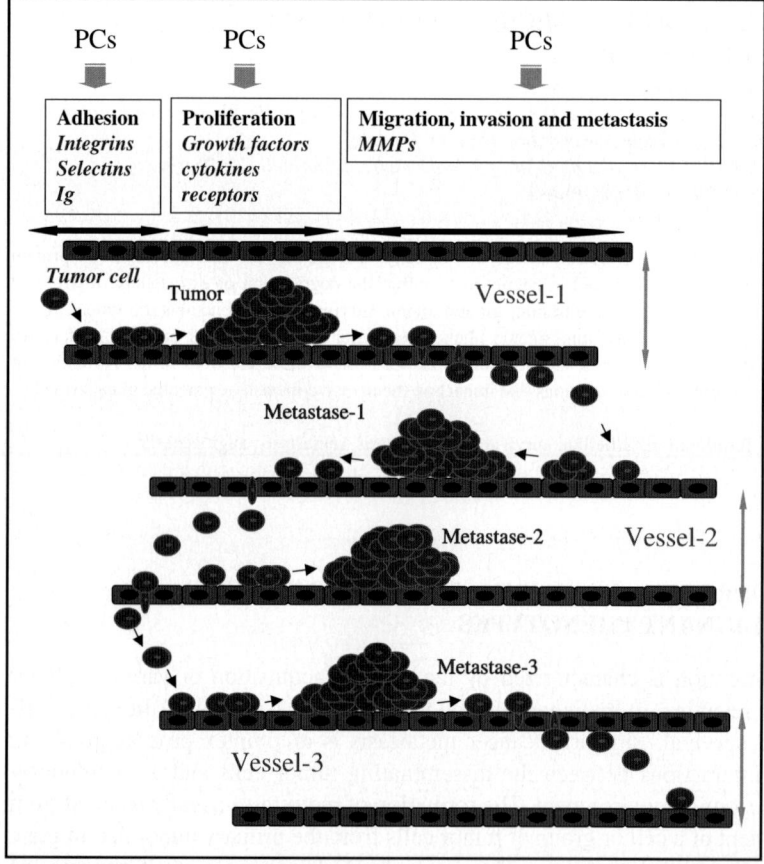

Figure 1. Cascade events implicating PCs for tumor growth and metastasis. PCs control tumor cell adhesion by activating or/and inducing the expression of adhesion molecules, regulate cell proliferation by activating growth factors, cytokines, and their receptors and induce migration and invasion of tumor cells that leads to metastases formation by activating MMPs

1.1 Role of the Proprotein Convertases in Tumor Cell Growth

Usually, non-transformed cells show an absolute requirement for growth factors for their proliferation in culture and generally more than one growth factor is required. The loss of/or decreased requirement for specific growth factors is a common occurrence in neoplastically transformed cells and may lead to a growth advantage, a fundamental feature of cancer cells. However *in vivo*, tumor development, growth, and progression depend on excessive growth factor pathway activation [1]. Many of these growth factors as synthesized as proproteins are processed and activated by PC-like enzymes [4] (Table 1). Thereby parallel increased expression of both growth factors and PCs may result in tumor growth advantage. Indeed, the anti-proliferate effect of α1-PDX, a general PC inhibitor on tumor cells was reported to be due to the inhibition of the processing and the function of various growth factors [5–8]. In initial, it was postulated that the critical nature of the Furin processing of various precursors may explain the antiproliferative effect of furin blockade on H-500 rat Leydig tumor cells [9], the pancreatic β-cell line MIN6 [10] and the gastric mucus cells [11]. Kayo et al. [10] showed that conditioned medium derived from furin-overexpressing MIN6 cells stimulated the growth of their parental control cells, whereas the medium derived from cells expressing α1-PDX resulted in a lower growth rate. These results suggest that high furin expression stimulates growth through an autocrine/paracrine mechanism. In agreement with this hypothesis, using directed mutagenesis assays, we found that the processing of PDGF-A [12] and VEGF-C [13] by the PCs is crucial for the mediation of tumorigenesis and tumor angiogenesis and lymphangiogenesis, respectively [12, 13]. Similarly, studies reported on TGF-β_1 processing by PCs revealed that inhibition of TGF-β_1 cleavage in glioblastoma and in HNSCC cell lines by α_1-PDX could be a promising tool in modulating tumor growth and immunotherapy [14]. Accordingly, over expression of Furin in various tumor cells were reported to increase the *in vitro* and *in vivo* growth rate of tumors by increasing probably the production of high amount of mature growth factors [15].

Like their ligands, several growth factor receptors contain also a consensus PC-cleavage site, and are critical for tumor growth and metastasis, e.g., the HGF receptor (HGF-R) [16], insulin receptor [17–19] and IGF-1 receptor [20] (Table 2).

Usually, receptor activation through ligand binding induces receptor dimerization and autophosphorylation through a *trans* mechanism. In turn, the transphosphorylation at specific tyrosine residues generates binding sites that are recognized by proteins involved in cellular signaling. Several proteins that are phosphorylated through growth factor-associated kinase activity were identified as downstream mediators of receptor-associated signal transduction. Overexpression, or mutations/insertions in growth factor receptors that result in constitutively high levels of proteins, or active kinases are documented in many tumors [21–25]. Frequently this is accompanied by constitutive high expression of the respective ligands providing an autocrine mechanism for growth autonomy [23–25]. Previously, the cleavage of insulin receptor was reported to be essential in the signal transduction of insulin [19]. In agreement with this observation, we previously

Table 1. Some growth factors with PC cleavage site (s)

Site (s) of processing		$P_6P_5P_4P_3P_2P_1{\downarrow}P'_1P'_2$	NCBI, Accession
Insulin	site-1	TPKTRREA	XP028180
	site-2	GSLQKRGI	
IGF-1		PAKSARSV	P01343
IGF-2		PAKSERDV	XP028189
PDGF-A		PIRRKRSI	NP002598
PDGF-B		LARGRRSL	NP148937
PDGF-C		FGRKSRVV	NP057289
PDGF-D		HDRKSKVD	AAK56136
VEGF-C		HSIIRRSL	P49767
VEGF-D		YSIIRRSI	NP_004460
TGF-β1		SSRHRRAL	XP_008912
Lefty protein		RSRGKRFS	O00292
BMP-2	site-1	HVRISRSL	
	site-2	HKREKRQA	NP_058874
BMP-4	site-1	HVRISRSL	
	Site-2	RRRAKRSP	NP_570912
BMP-7	site-1	HFRSIRST	NP_001710
Pancreatic polypeptide		PRYGKRHK	P01298
Gastrin		ASHHRRQ	P01350
GHRH	site-1	PFRMQRHV	
	site-2	RARLSRQE	NP_034415
FGF-23		PRRHTRSA	Q9GZV9
EGF	site-1	HHYSVRNS	P01133
	site-2	KWWELRHA	
Endothelin-1		LRRSKRCS	P05305
PTHRP		SRRLKRAV	P12272
Parathyroid hormone		KSVKKRSV	XP031173
Neurotrophin-3		TSRRKRYA	P20783
Neurotrophin-4		NRSRRGVS	A42687
β-NGF		THRSKRSS	XP002122
BDNF		SMRVRRHS	XP006027
APRIL		RSRKRRAV	O75888
BAFF		NSRNKRAV	Q9Y275
HB-EGF		RDRKVRDL	Q99075
HGF		KTKQLRVV	XP052260
LEAP-2		LCRKRRCS	NP_694709
Virokinin	site-1	LQRIARRP	
	site-2	GLMGKRDA	NM_013998

showed that the uncleaved IGF-1 receptor is unable to undergo the critical steps for IGF-1-mediated growth of tumor cells such as IGF-1-induced autophosphory-lation and IRS-1 phosphorylation [20]. However, Komada et al. previously reported that both cleaved and uncleaved HGF receptors can bind HGF and mediate their autophosphorylation and cell growth [16].

In several cases the inhibition of the PCs in tumor cells may affect the processing of both, the growth factor and its corresponding receptors. Such a situation was found

Table 2. Some growth factor receptors with PC cleavage site (s)

Site (s) of processing		$P_6P_5P_4P_3P_2P_1{\downarrow}P'_1P'_2$	NCBI, Accession
Insulin receptor		PSRKRRSL	XP048347
IGF-1 receptor		PERKRRDV	IGHUR1
HGF receptor		EKRKKRST	P08581
DREG		KIKVKRSL	NP_001027567
sorCS1	Site-1	SGRRRRSG	
	Site-2	ASRSPRGV	NP_443150
SorLA receptor		PLRRKRSA	U60975
Ldl-related protein		SNRHRRQI	Q07954
Leptin receptor		QVRGKRLD	P48357
Notch-1-receptor		SRKRRRQH	AAG33848
R-PTP-kappa		PHRTKREA	NP_002835

in the colon carcinoma cells line HT-29 that expresses IGF-1 receptor and its ligands IGF-1 and IGF-2 [26]. The expression of α_1-PDX in these tumor cells inhibited the processing of IGF-1R and probably its furin-processed ligands, IGF-1 and IGF-2, thereby abrogate their ability to maintain there growth through autocrine action [20].

1.2 Proprotein Convertases in Tumor Cells Survival

One of the hallmarks of various cancers is the deregulation of apoptotic or programmed cell death mechanisms usually found in normal cells that allow for corrupted cells to undergo cellular suicide. This includes mechanisms that attenuate pro apoptotic pathways and/or amplify anti apoptotic pathways. Increasing evidence suggests that many cancer cells use multiple and redundant pathways to maintain survival. Thereby the lost of mechanisms that controls the integrity of the cell erasing aberrant clones constitutes a key factor responsible for tumor formation and progression. Similarly to PCs role in cellular growth, their role in cell survival is also important. Previously, Lee R. et al found that the biological action of neurotrophins, growth factors that promote cell survival, differentiation, and cell death is regulated by their proteolytic cleavage [27]. Indeed, the nerve growth factor (NGF), when synthesized as proform is cleaved intracellularly by the furin and released under mature form. *In vitro* analysis revealed that the ProNGF is a high-affinity ligand for p75(NTR) and is able to induce with high affinity the p75NTR-dependent apoptosis processes in neurons. In contrast, following pro-NGF processing, the mature form is able to activate TrkA-mediated survival [27]. Thus processing of ProNGF controls both NGF-induced-cell survival and NGF-induced-cell death pathways. In our model, we found that the inhibition of the PCs activity in tumor cells by the expression of α1-PDX exaggerate the apoptotic phenotype induced by serum deprivation [20]. The role of PC in cellular survival can be explained by their role in the maturation of various proteins known to be anti-apoptotic mediators participating in various autocrine/paracrine mechanisms. A likely scenario may involve one or more secreted ligands and/or their receptors, all which require processing by PC-like enzymes. Examples include IGF-1 and IGF-2 that are synthesized and secreted by these cells [26] and processed by furin [4]. Since overexpression of α_1-PDX in tumor cells inhibited the processing of IGF-1R and its furin-processed ligands, IGF-1 and IGF-2, it is liable to abrogate their autocrine/paracrine protective effects.

The protective effect of many proteins is apparently dependent on their ability to induce a cascade of events leading to downstream pathways phosphorylation, including FAK, PI-3K1 and IRS-1 [28–30]. After phosphorylation these molecules mediate their anti-apoptotic effect through the activation of several negative death regulators such as Bcl-2 or inhibition of IL-converting enzyme (ICE)-like caspases [28–30]. Interestingly, expression of α_1-PDX in tumor cells resulted in a reduction of basal tyrosine phosphorylation of FAK, PI-3K and IRS-1 [20]. Exogenous addition of FAK, PI-3K and IRS-1 activators such as IGF-1, failed to increase tyrosine phosphorylation of these molecules in tumor cells and to rescue the cells from apoptosis. This suggests a blockade in the transmission of the autocrine/paracrine anti-apoptotic signals in these cells.

1.3   Regulation of Tumor Cell Invasion by the Proprotein Convertases

The breakdown of the barriers formed by extracellular matrix proteins is a pre-requisite for all processes of tissue remodeling. Matrix degradation reactions take part in physiological events but also represent a crucial step in cancer invasion. The role of the PCs in extracellular matrix degradation and the processes of invasion are now well documented [20, 31–37]. Inhibition of PCs in different tumor cells resulted in a significant reduction in their invasiveness [31]. This reduction is due to the processing blockade of proteins directly involved in the mechanism of invasion such as matrix metalloproteinases or to protein such as growth factors and/or integrins reported to induce the expression of ECM-degrading protein including urokinase and MMPs [31–37].

Recently, the inhibitory effect of α1-PDX on *in vivo* invasion was reported by Bassi et al. [31]. The *in vivo* invasion assay is based on the penetration of tumor cells into the wall of trachea implanted subcutaneously in mice. When transplanted in the tracheas the control-transfected cells penetrated deeply into the tracheal wall and reach the adventitia and the surrounding peritracheal tissues. Whereas, the penetration of α_1-PDX-transfected-cells in the tracheal wall was markedly decreased. We also found that α_1-PDX decreased the invasiveness of colon carcinoma HT-29 [20] in which the processing of MT1-MMP was inhibited [4, 31]. In addition, in these cells we found that the mRNA level of plasminogen activators urokinase-type plasminogen activator (uPA) and tissue-type plasminogen activator (tPA), the urokinase-type plasminogen activator receptor (uPAR), and the uPA inhibitor plasminogen activator inhibitor-1 (PAI-1), molecules involved in invasion processes and believed not to be processed by PCs, was significantly reduced [20]. This data demonstrated both the direct and the indirect involvement of PCs in the regulation of tumor cells invasion. The MMPs that were experimentally proven to be directly

Table 3. PC cleavage site (s) in MT-MMPs

Site (s) of processing		$P_6P_5P_4P_3P_2P_1 \downarrow P_1'P_2'$	NCBI, Accession
MT-1 MMP	site-1	AMR RPRC G	P50281
	site-2	NVR R K R YA	
MT-2 MMP	site-1	WMKRPR C G	P51511
	site-2	RR RRKRY A	
MT-3 MMP	site-1	WMK KP R CG	P51512
	site-2	HIR R K RY A	
MT-4 MMP	site-1	LMKTPR C S	Q9ULZ9
	site-2	QARRRR Q A	
MT-5 MMP	site-1	WMKKPRCG	Q9Y5R2
	site-2	RRRNKRYA	
MT-6-MMP		VRRRRRYA	NP_071913

involved in the processes of invasion and activated by the MMPs are Membrane-type matrix metalloproteases (MT-MMPs), Adamalysin metalloproteinases (ADAMs), Adamalysin metalloproteinases with thrombospondin motifs (ADAMTS) and Stromelysin-3. Others MMPS are only suspected to be PCs substrates based on the presence of a potential PCs cleavage sites in their amino acid sequences (Tables 3–7).

1.3.1 Membrane-type matrix metalloproteases (MT-MMPs)

Involved in metastasis, MT-MMPs were found to be overexpressed in a wide variety of carcinomas, including colon cancer [38]. These MMPs, particularly, MT1-MMP process proMMP-2 thereby enhancing invasiveness *in vitro* and *in vivo* as well [34, 47]. MMP-2 is the enzyme that degrades collagen 1V, a major type of collagen in basement membranes. MT1-MMP possesses two typical recognition motifs for PCs, namely ArgArgProArg[92] and ArgArgLysArg[111] that were reported to be cleaved by furin-like enzymes at both sites [34].

Analysis of MMPs amino acid sequences revealed the presence of potential cleavages sites for the PCs (Tables 4). However these observations required experiment validation as previously was done with the MT-MMPs and the other PCs substrates.

Table 4. Potential cleavage sites in several MMPs

Site (s) of processing	$P_6P_5P_4P_3P_2P_1 \downarrow P_1'P_2'$	NCBI, Accession
MMP-1	VMKQPRCG	P03956
MMP-2	TMRKPRCG	P08253
MMP-8	MMKKPRCG	XP006273
MMP-9	AMRTPRCG	XP029934
MMP-13	VMKKPRCG	XP040746
MMP-21	RARSRRSP	NP_671724
MMP-23	APRRRRYT	NP_004650
MMP-28	MRRKKRFA	NP_766385

To further understand the regulatory mechanisms that control the activity of this enzymes, using a soluble form of MT1-MMP, lacking the C-terminal transmembrane and cytoplasmic domains, Rozanov DV et al., found that active MT1-MMP was able to induce the activation of the zymogen and its self-proteolysis [47]. This autocatalytic processing generated six main forms of MT1-MMP and the N-terminal microsequencing of these fragments revealed that MT1-MMP functions as a self-convertase and is capable of cleaving its own prodomain at the furin cleavage site. This novel mechanism that control MT1-MMP processing may constitute an alternative pathways of MT1-MMP activation [47].

1.3.2   Adamalysin metalloproteinases

ADAMs are members of the membrane-associated multidomain zinc-dependent metalloproteinase family [48]. Sixteen of the thirty ADAM proteins identified to date are predicted to be catalytically active (Table 5), based on the presence of a conserved zinc binding sequence (HEXXH) in the protease domain, whereas the other members are not likely to be active proteases. The prodomains of several ADAM proteins such as ADAM12 are constitutively cleaved by a furin-type proprotein convertase [35].

This family was reported to play a role in diverse biological processes such as fertilization, myogenesis, neurogenesis and cell surface proteolysis and shedding of different proteins [35, 49–56]. Of the proteins found to be shed by the ADAMs, are cytokines and growth factors such as transforming growth factor-α (TGFα), epidermal growth factor (EGF), heparin-binding epidermal growth factor (HB-EGF), tumor necrosis factor-α (TNFα), c-Kit-ligand-1 (KL-1), colony-stimulating factor-1 (CSF-1), and Fas-ligand (Fas-L), receptors such as TNF receptor-I (TNFR1, p60 TNFR), TNF receptor-II (TNFR2, p80 TNFR), p75 nerve growth factor receptor (p75NGFR), interleukin 6 receptor (IL6R), thyroid-stimulating hormone receptor (TSHR), adhesion molecules such as L-selectin and others proteins such as protein tyrosine phosphatase σ(PTPσ), protein tyrosine phosphatase LAR (LAR), amyloid precurssor protein (APP) and angiotensin-converting enzyme (ACE).

Certain released precursor molecules can be cleaved by more than one enzyme, and some enzymes can cleave more than one substrate. For example, cleavage of TNF-α can be mediated by ADAM17 (TACE) and ADAM10, and α-secretase

Table 5. Some ADAMs with PC cleavage site (s)

Site (s) of processing	$P_6P_5P_4P_3P_2P_1\!\downarrow\!P_1'P_2'$	NCBI, Accession
ADAM1	PPRSRKPD	AAA74920
ADAM8	PSRETRYV	XP005675
ADAM9	LLRRRRAV	NP003807
ADAM10	LLRKKRTT	XP007741
ADAM12	ARRHKRET	XP005838
ADAM15	HIRRRRDV	Q13444
ADAM17	VHRVKRRA	P78536

activity for amyloid precursor protein has been attributed also to ADAM17 [51], ADAM9 [54] and ADAM10 [56]. It is not known how these protease(s) select their substrate, because reliable consensus cleavage sites have not been identified. In addition to their proteolytic function, some members of the ADAM family like ADAM15 and ADAM2 can support integrin binding via their disintegrin domain [55, 57, 58]. It is increasingly recognized that ADAMs represent a novel group of membrane proteases that are important for cellular interactions under physiological and pathophysiological conditions including cancer. Recently, several ADAM family members were described to be dramatically up-regulated in many tumor cells. This includes cells derived from a range of hematological malignancies [49] and breast, prostate, lung and colon cancer [36, 48, 62]. Therefore, it is of particular interest that the member of the ADAM family reported to shed cell-associated neural adhesion molecules such as L1 may be relevant to promote cell migration and invasion [62, 63]. Interestingly, in these cells the putative tumor suppressor gene MDC (ADAM11) was expressed at very low level [59, 60].

1.3.3   Adamalysin metalloproteinases with thrombospondin motifs

ADAM-TSs are members of the ADAM family containing thrombospondin type motifs. These enzymes have been implicated in various cellular events, including cleavage of proteoglycans, extracellular matrix degradation, inhibition of angiogenesis, gonadal development, and organogenesis. The first member of this family, called ADAM-TS1, was originally cloned from colon adenocarcinoma cell line [36, 37]. Based on its capacity to form a covalent complex with $\alpha 2$-macroglobulin, recent studies demonstrated that ADAM-TS1 protein is proteolytically active [36, 37] (Table 6). In addition, the maturation of ADAM TS1 precursor is impaired in the furin-deficient cell line, LoVo and the processing ability of the cells is restored by the co-expression of the furin cDNA [61].

The only members of the ADAM-TS family with established substrates are ADAM-TS2 (procollagen-N-proteinase) and ADAM-TS4 and -TS11 (aggrecanases-1 and -2, respectively) [61, 62]. These proteinases cleave aggrecan at one or more of five specific sites in the aggrecan core protein [36, 61–63]. Previously, the expression of ADAM-TS4 was reported to be induced during endothelial cells undergoing differentiation into tube like structures suggesting its implication in

Table 6. Some ADAMTS with PC cleavage site (s)

Site (s) of processing		$P_6P_5P_4P_3P_2P_1^{\downarrow}P_1'P_2'$	NCBI, Accession
ADAMTS-1		SIRKKRFV	Q9UHI8
ADAMTS-2	site-1	GVRTRRAA	P79331
	site-2	RRRMRRHA	
ADAMTS-3		TMRRRRHA	O15072
ADAMTS-4		PRRAKRFA	XP042446
ADAMTS-5/11		WRRRRRS I	Q9UNA0
ADAMTS-13		RQRQRRAA	CAC83682

Table 7. Stromelysins with PC cleavage site (s)

Site (s) of processing		$P_6P_5P_4P_3P_2P_1^{\downarrow}P_1'P_2'$	NCBI, Accession
STR-1		VMRKP R CG	XM 058067
STR-2		VMRKPRCG	AAH02591
STR-3	site-1	SLRPPRCG	P24347
	site-2	RNRQKRFV	

angiogenesis [64]. These observations link ADAM-TS family members to invasion and the blocking the activation of ADAM-TSs by PC inhibitors may provide a novel therapeutic strategy.

1.3.4   Stromelysin-3 (Str-3)

Stromelysin 3 (Str-3 or MMP 11) is one of the first members of the MMP family that found to be activated by furin-like activity (Table 7). The processing Str-3 occurred intracellularly before its secretion [66].

Previously, Str-3 expression has been tightly linked with the extent of local invasiveness of various cancers, including breast, head and neck, as well as basal skin cancers, implicating its important function in multiple epithelial malignancies [65, 67–70]. In the case of breast carcinomas, elevated Str-3 mRNA levels in the primary tumors are highly predictive for the presence of distant metastases [71]. Str-3 displays a distinct pattern of substrate specificity which includes serine protease inhibitors (serpins) such as α1-antitrypsin) [72]. Enzymatic degradation of this inhibitor by Str-3 seems of particular interest because α1-antitrypsin has been shown to be a potent inhibitor of anchorage independent growth in epithelial carcinoma cells [73]. Furthermore, α1-antitrypsin functions specifically as an inhibitor of leucocyte elastase, a tissue destructive proteinase stored in neutrophils and monocytes [74]. Taking these observations together, these studies suggest that furin dependent activation of Str-3 its self dependent inactivation of α1-antitrypsin simultaneously affects the proliferative and invasive activity of tumor cells.

1.4   Regulation of Tumor Cell Adhesion by the Proprotein Convertases

Cell adhesion molecules control cellular traffic, transmigration through the endothelium, homing in and localization to various target organs during inflammation and tumor cell colonization [75, 76]. Most of theses molecules characterized so far fall into Integrin, Immunoglobulin, Selectin or cadherin families.

1.4.1   The integrin family

Integrins are the most important receptors for ECM proteins, such as fibronectin, laminins, collagens or vitronectin. These heterodimeric cell surface adhesion receptors are formed by two noncovalently associated subunits, α and β. There are 18 α and 8 β subunits which associate to form 24 different heterodimers [77, 78].

Table 8. Some Integrins with PC cleavage site (s)

Site (s) of processing	$P_6P_5P_4P_3P_2P_1{\downarrow}P_1'P_2'$	NCBI, Accession
Integrin α3	PQRRRQL	XP008432
Integrin α4	HVISKRST	XP039011
Integrin α5	HHQQKREA	AAH08786
Integrin α6	NSRKKREI	NP000201
Integrin α7	RDRRRREL	Q13683
Integrin α8	HLVRKRDV	AAA93514
Integrin α^E	TARQRRAL	XP008508
Integrin αIIb	HKRDRRQI	P08514
Integrin αv	HLITKRDL	XP002379

They bind ligands such as the ICAMs, VCAMs and several ECM components. Integrin subunits consist of a large extracellular domain, a single transmembrane domain and a short cytoplasmic tail. Integrins are promiscuous and redundant receptors: i.e. one integrin can bind several different ligands, and many different integrins can bind to the same ligand. The most promiscuous integrin is αvβ3 which can bind to over a dozen of different ligands, including fibronectin, vitronectin, thrombospondin, tenascin-C, fibrin, von Willenbrand Factor, osteo-pontin and denatured collagen I [78, 80]. The extracellular domains of both subunits are required for ligand binding while the cytoplasmic tails interact with the cytoskeleton, induce changes in cell shape and motility and transduce growth and survival signals [81, 82]. Previously, altered expression, or activation of integrins was linked to various cancers. Activation of integrins was reported to mediate MMPs and urokinase activity of many tumor cells, including melanoma and colon carcinoma [84–86] (Table 8).

While β-subunits are not cleaved by PCs, a total of 9 out of 18 known α–subunits possess a potential PC-cleavage site, with α3, α4, α5, α6 and αv proven to be PC5 and furin substrates [87, 88].

Previously, Berthet et al. reported that the endoproteolytic cleavage of αv integrin is important for the signal transduction pathway mediating cell adhesion [88]. Blockade of αvβ5 cleavage resulted in a decreased phosphorylation of focal adhesion kinase (FAK) and paxillin, two important molecules involved in cell adhesion. Indeed, tumor cells with uncleaved αvβ5 integrin revealed with altered attachment to vitronectin.

1.4.2   The Immunoglobulin family

The Ig-family includes intercellular adhesion molecules ICAM-1, ICAM-2, and ICAM-3, vascular cell adhesion molecule-1 (VCAM-1), and mucosal addressin cell adhesion molecule-1 (MadCAM-1), none of which are believed to be substrates for PC-processing. However, the convertases seem to be required for their expression and probably function. Indeed, the expression of ICAM-1 and VCAM-1 on endothelial cells is induced by various cytokines and growth factors such as interferon-gamma (INF-γ), interleukin-1 (IL-1), and tumor necrosis factor (TNF-α) insulin-like growth factor (IGF-1), endothelins [89–97]. Some of these precursors were reported to be directly processed by PCs such as IGF-1 and endothelins [5, 97]. Others are indirectly regulated by the PCs through the cleavage of their cognate enzymes or inducers, for example tumor necrosis factor-alpha converting enzyme (TACE) [51] and ADAM10 [53] that process proTNF-α are themselves activated by PCs. These adhesion molecules were reported to confer metastatic potential and increased tumorigenicity to various tumor cells. They mediate homotypic and heterotypic adhesion between tumor cells and endothelial cells, respectively. In addition both types of interaction were reported to promote metastasis at different stages in the metastasis cascade.

1.4.3   The selectin family

The selectin family members, L-selectin, P-selectin and E-Selectin were reported to be involved in the adhesion of leukocytes to activate the endothelium. Adhesion is initiated by weak interactions that produce a characteristic "rolling" motion of leukocytes on the endothelial surface. This rapid "on-off" attachment is necessary for activation and engagement of integrins and their counter-receptors on the leukocytes and endothelial cells, respectively. The integrin-mediated, high-affinity binding is in turn required for leukocyte arrest and extravasation [75]. On the basis of *in vitro* studies it is postulated that similar cell-cell interaction may also occur between circulating malignant cells and the vascular endothelium during tumor dissemination [42, 99, 100]. In general, the selectins bind sialytated, glycosylated or sulfated glycans on glycoproteins, glycolipids or proteoglycans [101]. The tetrasaccharides sialyl-Lewis^x (sLew^x) and sialyl-Lewis^a (sLew^a) are recognized by all three selectins. *In vitro* adhesion studies showed that human colorectal, pancreatic and gastric carcinoma cells utilize sLew^x and related carbohydrate determinants to adhere to TNF-α-inducible E-selectin on cultured vascular endothelial cells [40, 42, 44, 58, 91, 102, 103]. *In vivo* studies in turn showed that inhibition of liver metastasis of the highly metastatic PCI cells could be blocked with antibodies to sLew^a and that lung colonization by colon carcinoma HT29 cells could be blocked by a soluble E-selectin fusion protein [102, 103].

Under normal physiological conditions, vascular endothelial cells express low constitutive levels of E-selectin. Several cytokines and growth factors such as IL-1, TNF-α, vascular endothelial growth factor (VEGF), IGF-1 and endothelins induce the expression of E-selectin [5, 96]. Many of these molecules are directly or indirectly require PC-activity. Thus, although like VCAMs and ICAMs, except L selectin (Table 9), these molecules are not processed by PCs. The latter may indirectly control their expression *via* the activation of some of the above inducers. Recently, we demonstrated that E-selectin expression on sinusoids was described to be one of the early molecular events involved in liver metastasis [40, 44]. The arrest of tumor cells in the hepatic circulation causes a cascade of events, which start with a rapid release of IL-1 and TNF-α (and other mediators). In turn, these

Table 9. PC cleavage site in L1 CAM

Site (s) of processing	$P_6P_5P_4P_3P_2P_1^\downarrow P_1'P_2'$	NCBI, Accession
L1 CAM	RKHSKRHI	NP_000416

stimulate the expression of E-selectin on hepatic endothelial cells, resulting in an enhanced tumor cell adhesion and subsequent metastases [40, 44].

1.4.4   Cadherins

More than 80% of human cancers are originated from epithelial cells. In normal tissues, epithelial cells are tightly interconnected through several junctional structures, including tight junctions, adherens-type junctions and desmosomes, which are intimately associated with the actin and intermediate cytoskeleton [107].

The adherens are constituted with the transmembrane glycoproteins cadherins and desmosomes that are constitute with desmogleins and desmocollins. The epithelial cadherin (E-cadherin) is the major cadherin present in adherens junctions. E-cadherin is linked to actin cytoskeleton trough its interaction with the catenins [α, β and γ catenin (also known as plakoglobin). During the development of epithelial cancers, such as those from, colon, stomach, liver, esophagus, skin, lung and breast the adhesion function of E-cadherin is frequently lost and correlates with tumor growth and metastasis [108, 109]. Similarly, other members of the cadherin family are also expressed in epithelia, but very few have been implicated in cancer. For example, H-cadherin expression is lost during the development of breast cancer [110]. In contrast N-cadherin was reported to enhance the invasiveness of various cancer cells [111], and high levels of cadherin-6 were associated with metastasis [112] and suggested as potential marker for circulating cancer cells in renal cell carcinoma [113]. These divergences in the role of cadherins in tumor progression and metastasis raised the question whether the loss or increase of cadherins are a pre-requisite for tumor progression, or whether it is a consequence of tumor progression and metastasis.

E-cadherin like other cadherin members family is synthesized as protein precursors that required proteolytic cleavage by the PCs (Table 10) [114, 115].

Recently, Posthaus et al. reported that the E-cadherin precursor processing in mouse keratinocytes was not completely abrogated by α1-PDX. This inhibitor was reported to potently inhibit furin and PC5 and to a lesser degree PACE4, and had not effect on PC7 [115]. These results are consistent with recent findings indicating partial inhibition of proE-cadherin cleavage in α1-PDX-transfected Lovo cells [116]. The authors of this study suggested that the procadherin processing relies on cleavage by the convertases furin, PACE4 and/or PC5 and when these convertases are inhibited in LoVo cells by α1-PDX expression, the α1-PDX-insensitive convertase PC7 mediate the processing of procadherin. The processing of E-cadherin was reported to be crucial for the acquisition of its function and act as tumor suppressors [114].

Table 10. PC cleavage site in Cadherins

Site (s) of processing	$P_6P_5P_4P_3P_2P_1{\downarrow}P_1'P_2'$	NCBI, Accession
Cadherin 4	LRRRKRDW	NM_001794
Cadherin 5	LQRHKRDW	NP_001003983
Cadherin-6	LSRSKRSW	NP_037059
Cadherin-7	RSRTKRSW	CAC13127
Cadherin-8	LNRSKRGW	D38992
Cadherin-9	LHRAKRGW	XP_487374
Cadherin-10	LHRQKRGW	NP_006718
Cadherin-11	LQRSKRGW	I49556
Cadherin-12	FQRVKRGW	P55289
Cadherin-13	PRQKRSIV	Q9WTR5
Cadherin-14	HHRPKRGW	AAB0293
Cadherin-15	SRVRRAWV	NP_004924
Cadherin-18	HHRPKRGW	AAH31051
Cadherin-20	QRTKRSWV	Q9HBT6
Cadherin-22	AGRVKRGW	NM_019161
E-Cadherin	LRRQKRDW	BAA88957
BR-Cadherin	FQRVKRGW	P55289
C-Cadherin	LKRKKRDW	AAC16910
EP-Cadherin	LKRKKRDW	IJXLCP
F-Cadherin	HHRLKRSW	I51638
H-Cadherin	VPRQKRSI	NM_001257
N-Cadherin	ALQRQKRD	Q9Z1Y3
OB-Cadherin	LQRSKRGW	BAA04798
P-Cadhedin	LRRRKREW	P10287

2.   PROPROTEIN CONVERTASES IN TUMORIGENESIS, ANGIOGENESIS AND METASTASIS

The antitumor effect of PC inhibition was initially reported by our group. Expression of α1-PDX in colon carcinoma HT-29 tumor cells delayed the appearance, the incidence and the vascularization of palpable tumors [20]. Subsequently, many studies have confirmed the inhibition of tumor growth by PC inhibition [31]. Accordingly, elevated expression of different PCs was reported for different human cancers and the levels of the proprotein convertase furin mRNA and protein expression correlate with the aggressiveness of various tumor cell lines [15, 31]. Interestingly, analysis of tumors derived from tumor cells expressing the PC inhibitor α1-PDX-injected mice, revealed the existence of significant loss in α_1-PDX expression in the subcutaneous tumors during their growth [31]. This observation indicates that the inhibitory effect of α_1-PDX on tumor growth and invasiveness in *in vivo* systems may be underestimated. However, it is not clear whether the observed progressively lower levels of α1-PDX mRNA in tumors is due to a specific loss of the cDNA from tumor cells. Similar observations were reported by Leitlein et al. [14] on the loss of α1-PDX expression *ex vivo*.

Metastasis is a multistep process during which cancer cells disseminate from the site of primary tumors and establish secondary tumors in distant organs. Several lines of evidence indicate that in the first step of metastatic invasion, tumor cells interact with endothelial cells through the use of several distinct adhesion molecules[39–44]. e.g., in the liver, the arrest of tumor cells in the hepatic circulation induce various cytokines such as IL-1 and TNF-α that in turn up-regulate the endothelial expression of various adhesion molecules, particularly E-selectin [39–44]. The latter was described as a crucial adhesion molecule in tumor metastasis as revealed by the metastatic spread to the liver of tumor cells that was inhibited by an anti-E-selectin antibody [43]. Accordingly, we previously found that TNF-α mRNA expression was very rapidly augmented in the liver within 6 h after intrasplenic tumor injection [40, 44]. The expression of this cytokine was also found to be enhanced around central and portal veins, presumed to be the entry site of tumor cells to the liver [41]. Thus, rapidly migrating tumor cells that may induce the expressions of IL-1, TNF-α and other mediators rapidly enhance the expression of E-selectin, thereby increasing the interaction between tumor and endothelium cells, the first step of liver tumor colonization.

Using two colon cancer cell line HT-29 and CT-26 that preferentially metastasize to the liver, we found that the expression of α1-PDX in these cells inhibit their ability to make metastasis in the liver (unpublished data). These cells were unable to induce cytokines and E-selectin expression in liver when injected into mice hepatic circulation (unpublished data). These data indicate that the PCs contribute to metastasis by promoting active molecules involved in the induction of the first step of liver colonization by tumor cells. Accordingly, using internally quenched synthetic fluorogenic peptide mimicking the cleavage site of PC found in proVEGF-C [13], we found that the ability of tumor cells expressing α1-PDX-derived metastatic livers to process VEGF-C is significantly decreased, thereby confirming the reduced activity of PCs in tumors derived from mice injected with tumor cells that express α1-PDX.

Recently, using immunosuppressed newborn rats, Nejjari M et al., confirmed the importance of the PCs in tumorigenicity of human colon carcinoma HT-29 as revealed by the reduced size of subcutaneous tumors in response to α1-PDX expression [45]. In contrast they found that the inhibition of the PCs increases the number of lung metastasis in their animal model. These data suggest that by processing and activating a wide range of protein precursors, the PCs could be considered as pleiotropic molecules; therefore, it is not surprising that PCs may have diverse effects on tumor progression and metastasis. In addition to the difference of the animal model used in Nejjari M et al study, the apparently opposing role of PCs in lung and liver metastasis could be attributed to many other factors: this include the difference in the adhesion of metastatic cells to the endothelium of the host organ which is of major importance in initiating the arrest of tumor cells prior to invasion, the difference in metastatic potential of the same cells in different organs, and finally, the difference in the released agent(s) during detachment of a cell from a primary tumor or attachment to the host organ can also provide an explanation.

Angiogenesis and the development of tumor and metastases are intrinsically connected. Although angiogenesis is a significant prognostic factor in various cancers, the factors that control this process *in vivo* are not well defined. Usually tumor angiogenesis is mediated by tumor-secreted angiogenic growth factors that interact with their surface receptors expressed on endothelial cells. Multiple angiogenic proteins are known, including VEGF and its four isoforms *[121, 165, 189, and 206* amino acids], TGF-β1, pleiotrophin acidic and bFGF. Previously, immunohistochemical analysis of CD31 antigen expression, a marker of endothelial cells revealed a reduced tumor vascularization of tumors developed from tumor cells expressing α1–PDX [4, 20]. Using site-directed mutagenesis to alter the PC recognition motif in pro-VEGF-C in order to study directly the importance of VEGF-C processing in the mediation of its cellular function and tumorigenesis, we found that unprocessed VEGF-C lost its ability to induce angiogenesis and lymphangiogenesis [13]. This suggests the importance of the PCs in tumor vessel formation through direct/indirect activation of various angiogenic proteins.

REFERENCES

[1] Aaronson SA (1991) Growth factors and cancer. Science **254**:1146–1153
[2] Schwartz MA (1997) Integrins, oncogenes, and anchorage independence. J Cell Biol. **139**:575–578
[3] DeClerck YA (2000) Interactions between tumour cells and stromal cells proteolytic modification of the extracellular matrix by metalloproteinases in cancer. Eur J Cancer **36**:1258–1268
[4] Khatib AM, Siegfried G, Chrétien M, Metrakos P, Seidah NG (2002) Proprotein convertases in tumor progression and malignancy: Novel targets in cancer therapy. Am J Pathol. (1995) **160**:1921 1935
[5] Duguay SJ, Lai-Zhang J, Steiner DF (1995) Mutational analysis of the insulin-like growth factor I prohormone processing site. J Biol Chem **270**:17566–17574
[6] Duguay SJ, Jin Y, Stein J, Duguay AN, Gardner P, Steiner DF (1998) Post-translational processing of the insulin-like growth factor-2 precursor: Analysis of O-glycosylation and endoproteolysis. J Biol Chem **273**:18443–18451
[7] Duguay, SJ (1991) Post-translational processing of insulin-like growth factor. Horm Metab Res **31**:43–49
[8] Campan M, Yoshizumi M, Seidah NG, Lee ME, Bianchi C, Haber E (1996) Increased proteolytic processing of protein tyrosine phosphatase mu in confluent vascular endothelial cells: The role of PC5, a member of the subtilisin family. Biochemistry **35**:3797–3802
[9] Liu B, Amizuka N, Goltzman D, Rabbani SA (1995) Inhibition of processing of parathyroid hormone-related peptide by anti-sense furin: Effect in vitro and in vivo on rat Leydig (H-500) tumor cells. Int J Cancer **63**:276–281
[10] Kayo T, Sawada Y, Suda M, Konda Y, Izumi T, Tanaka S, Shibata H, Takeuchi T (1997) Proprotein-processing endoprotease furin controls growth of pancreatic beta-cells. Diabetes **46**:1296–1304
[11] Konda Y, Yokota H, Kayo T, Horiuchi T, Sugiyama N, Tanaka S, Takata K, Takeuchi T (1997) Proprotein-processing endoprotease furin controls the growth and differentiation of gastric surface mucous cells. J Clin Invest **99**:1842–1851
[12] Siegfried G, Khatib AM, Benjannet S, Chretien M, Seidah NG (2003) The proteolytic processing of pro-platelet-derived growth factor-A at RRKR(86) by members of the proprotein convertase family is functionally correlated to platelet-derived growth factor-A-induced functions and tumorigenicity. Cancer Res **63**:1458–1463

[13] Siegfried G, Basak A, Cromlish JA, Benjannet S, Marcinkiewicz J, Chretien M, Seidah NG, Khatib AM (2003) The secretory proprotein convertases furin, PC5, and PC7 activate VEGF-C to induce tumorigenesis. J Clin Invest. **111**:1723–1732

[14] Leitlein J, Aulwurm S, Waltereit R (2001) Processing of immunosuppressive pro-TGF-beta-1,2 by human glioblastoma cells involves cytoplasmic and secreted furin-like proteases. J Immunol **166**:7238–7243

[15] Bassi DE, Mahloogi H, Lopez De Cicco R, Klein-Szanto A (2003) Increased furin activity enhances the malignant phenotype of human head and neck cancer cells. Am J Pathol **162**:439–447

[16] Komada M, Hatsuzawa K, Shibamoto S, Ito F, Nakayama K, Kitamura N (1993) Proteolytic processing of the hepatocyte growth factor/scatter factor receptor by furin. FEBS Lett **328**:25–29

[17] Papa V, Pezzino V, Costantino A, Belfiore A, Giuffrida D, Frittitta L, Vannelli GB, Brand R, Goldfine ID, Vigneri R (1990) Elevated insulin receptor content in human breast cancer. J Clin Invest **86**:1503–1510

[18] Robertson BJ, Moehring JM, Moehring TJ (1993) Defective processing of the insulin receptor in an endoprotease-deficient Chinese hamster cell strain is corrected by expression of mouse furin. J Biol Chem **268**:24274–24277

[19] Hwang JB, Hernandez J, Leduc R, Frost SC (2000) Alternative glycosylation of the insulin receptor prevents oligomerization and acquisition of insulin-dependent tyrosine kinase activity. Biochem Biophys Acta **1499**:74–84

[20] Khatib AM, Siegfried G, Prat A, Luis J, Chretien M, Metrakos P, Seidah NG (2001) Inhibition of proprotein convertases is associated with loss of growth and tumorigenicity of HT-29 human colon carcinoma cells: Importance of insulin-like growth factor-1 (IGF-1) receptor processing in IGF-1-mediated functions. J Biol Chem **276**:30686–30693

[21] Baserga R, Rubin R (1993) Cell cycle and growth control. Crit Rev Eukaryot Gene Expr **3**:47–61

[22] Rodrigues GA, Park M (1994) Oncogenic activation of tyrosine kinases. Curr Opin Genet **4**:15–24

[23] Rodrigues GA, Park M (1994) Autophosphorylation modulates the kinase activity and oncogenic potential of the Met receptor tyrosine kinase. Oncogene **9**:2019–2027

[24] Werb Z, Tremble PM, Behrendtsen O, Crowley E, Damsky CH (1989) Signal transduction through the fibronectin receptor induces collagenase and stromelysin gene expression. J Cell Biol **109**:877–889

[25] Hiscox SE, Hallett MB, Puntis MC, Nakamura T, Jiang WG (1997) Expression of the HGF/SF cancers. Cancer Invest **15**:513–521

[26] Lahm H, Amstad P, Wyniger J, Yilmaz A, Fischer JR, Schreyer M, Givel JC (1994) Blockade of the insulin-like growth-factor-I receptor inhibits growth of human colorectal cancer cells: Evidence of a functional IGF-II-mediated autocrine loop. Int J Cancer **58**:452–459

[27] Lee R, Kermani P, Teng KK, Hempstead BL (2001) Regulation of cell survival by secreted proneurotrophins. Science **294**:1945–1948

[28] Franke TF, Kaplan DR, Cantley LC (1997) PI3K: Downstream AKTion blocks apoptosis. Cell Feb **88**:435–437

[29] Ueno H, Kondo E, Yamamoto-Honda R, Tobe K, Nakamoto T, Sasaki K, Mitani K, Furusaka A, Tanaka T, Tsujimoto Y, Kadowaki T, Hirai H (2000) Association of insulin receptor substrate proteins with Bcl-2 and their effects on its phosphorylation and antiapoptotic function. Mol Biol Cell **11**:735–746

[30] Jarpe MB, Widmann C, Knall C, Schlesinger TK, Gibson S, Yujiri T, Fanger GR, Gelfand EW, Johnson GL (1998) Anti-apoptotic versus pro-apoptotic signal transduction: Checkpoints and stop signs along the road to death. Oncogene **17**:1475–1482

[31] Bassi DE, Lopez De Cicco R, Mahloogi H, Zucker S, Thomas G, Klein-Szanto AJ (2001) Furin inhibition results in absent or decreased invasiveness and tumorigenicity of human cancer cells. Proc Natl Acad Sci USA **98**:10326–10331

[32] Santavicca M, Noel A, Angliker H, Stoll I, Segain JP, Anglard P, Chretien M, Seidah N, Basset P (1996) Characterization of structural determinants and molecular mechanisms involved in pro-stromelysin-3 activation by 4-aminophenylmercuric acetate and furin-type convertases. Biochem J **315**:953–958

[33] Pei D, Weiss SJ (1995) Furin-dependent intracellular activation of the human stromelysin-3 zymogen. Nature **375**:244–247

[34] Yana I, Weiss SJ (2000) Regulation of membrane type-1 matrix metalloproteinase activation by proprotein convertases. Mol Biol Cell **11**:2387–2401

[35] Loechel F, Gilpin BJ, Engvall E, Albrechtsen R, Wewer UM (1998) Human ADAM 12 (meltrin α) is an active metalloprotease. J Biol Chem **273**:16993–16997

[36] Kuno K, Kanada N, Nakashima E, Fujiki F, Ichimura F, Matsushima K (1997) Molecular cloning of a gene encoding a new type of metalloproteinase-disintegrin family protein with thrombospondin motifs as an inflammation associated gene. J Biol Chem **272**:556–562

[37] Kuno K, Terashima Y, Matsushima K (1999) ADAMTS-1 is an active metalloproteinase associated with the extracellular matrix. J Biol Chem **274**:18821–18826

[38] Sardinha TC, Nogueras JJ, Xiong H, Weiss EG, Wexner SD, Abramson S (2000) Membrane-type 1 matrix metalloproteinase mRNA expression in colorectal cancer. Dis Colon Rectum **43**:389–395

[39] Kitakata H, Nemoto-Sasaki Y, Takahashi Y, Kondo T, Mai M, Mukaida N (2002) Essential roles of tumor necrosis factor receptor p55 in liver metastasis of intrasplenic administration of colon 26 cells. Cancer Res **62**:6682–6687

[40] Khatib AM, Auguste P, Fallavollita L, Wang N, Samani A, Kontogiannea M, Meterissian S, Brodt P (2005) Characterization of the host proinflammatory response to tumor cells during the initial stages of liver metastasis. Am J Pathol **167**:749–759

[41] Kitakata H, Nemoto-Sasaki Y, Takahashi Y, Kondo T, Mai M, Mukaida N (2002) Essential roles of tumor necrosis factor receptor p55 in liver metastasis of intrasplenic administration of colon 26 cells. Cancer Res. **62**:6682–6687

[42] Tozeren A, Kleinman H.K, Grant D.S, Morales D, Mercurio A.M, Byers S.W (1995) E-selectin-mediated dynamic interactions of breast- and colon-cancer cells with endothelial-cell monolayers. Int J Cancer **60**:426–431

[43] Brodt P, Fallavollita L, Bresalier R.S, Meterissian S, Norton C.R, Wolitzky B.A (1997) Liver endothelial E-selectin mediates carcinoma cell adhesion and promotes liver metastasis. Int J Cancer **71**:612–619

[44] Khatib AM, Fallavollita L, Wancewicz EV, Monia BP, Brodt P (2002) Inhibition of hepatic endothelial E-selectin expression by C-raf antisense oligonucleotides blocks colorectal carcinoma liver metastasis. Cancer Res **62**:5393–5398

[45] Nejjari M, Berthet V, Rigot V, Laforest S, Jacquier MF, Seidah NG, Remy L, Bruyneel E, Scoazec JY, Marvaldi J, Luis J (2004) Inhibition of proprotein convertases enhances cell migration and metastases development of human colon carcinoma HT-29 cells in a rat model. Am J Pathol **164**:1925–1933

[46] Murakami K, Sakukawa R, Ikeda T, Matsuura T, Hasumura S, Nagamori S, Yamada Y, Saiki I (1999) Invasiveness of hepatocellular carcinoma cell lines: Contribution of membrane-type 1 matrix metalloproteinase. Neoplasia **1**:424–430

[47] Rozanov DV, Strongin AY (2003) Membrane type-1 matrix metalloproteinase functions as a proprotein self-convertase. Expression of the latent zymogen in Pichia pastoris, autolytic activation, and the peptide sequence of the cleavage forms. J Biol Chem **278**:8257–8260

[48] Primakoff P, Myles DG (2000) The ADAM gene family: Surface proteins with adhesion and protease activity. Trends Genet **16**:83–87

[49] Wu E, Croucher PI, McKie N (1997) Expression of members of the novel membrane linked metalloproteinase family ADAM in cells derived from a range of haematological malignancies. Biochem Biophys Res Commun. **235**:437–442

[50] Schlöndorff J, Blobel CP (1999) Metalloprotease-disintegrins: Modular proteins capable of promoting cell-cell interactions and triggering signals by protein-ectodomain shedding. J Cell Sci **112**:3603–3617

[51] Black RA, White JM (1998) ADAMs: Focus on the protease domain. Curr Opin Cell Biol **10**:564–569

[52] Wolfsberg TG, White JM (1996) ADAMs in fertilization and development. Dev Biol **180**:389–401

[53] Lunn CA, Fan X, Dalie B, Miller K, Zavodny PJ, Narula SK, Lundell D (1997) Purification of ADAM 10 from bovine spleen as a TNF-α convertase. FEBS Lett **400**:333–335

[54] Koike H, Tomioka S, Sorimachi H, Saido TC, Maruyama K, Okuyama A, Fujisawa-Sehara A, Ohno S, Suzuki K, Ishiura S (1999) Membrane-anchored metalloprotease MDC9 has an alpha-secretase activity responsible for processing the amyloid precursor protein. Biochem J **343**:371–375

[55] Nath D, Slocombe PM, Stephens PE, Warn A, Hutchinson GR, Yamada KM, Docherty AJ, Murphy G (1999) Interaction of metargidin (ADAM-15) with alphavbeta3 and alpha5beta1 integrins on different haemopoietic cells. J Cell Sci **112**:579–587

[56] Lopez-Perez E, Zhang Y, Frank SJ, Creemers J, Seidah N, Checler F (2001) Constitutive alpha-secretase cleavage of the beta-amyloid precursor protein in the furin-deficient LoVo cell line: Involvement of the pro-hormone convertase 7 and the disintegrin metalloprotease ADAM10. J Neurochem **76**:1532–1539

[57] Chen MS, Almeida EA, Huovila AP, Takahashi Y, Shaw LM, Mercurio AM, White JM (1999) Evidence that distinct states of the integrin alpha6beta1 interact with laminin and an ADAM. J Cell Biol **144**:549–561

[58] Eto K, Puzon-McLaughlin W, Sheppard D, Sehara-Fujisawa A, Zhang XP, Takada Y (2000) RGD-independent binding of integrin alpha 9beta 1 to the ADAM-12 and -15 disintegrin domains mediates cell-cell interaction. J Biol Chem **275**:34922–34930

[59] Gutwein P, Oleszewski M, Mechtersheimer S, Agmon-Levin N, Krauss K, Altevogt P (2000) Role of Src kinases in the ADAM-mediated release of L1 adhesion molecule from human tumor cells. J Biol Chem **275**:15490–15497

[60] McCulloch DR, Harvey M, Herington AC (2000) The expression of the ADAMs proteases in prostate cancer cell lines and their regulation by dihydrotestosterone. Mol Cell Endocrinol **167**:11–21

[61] Tortorella M, Pratta M, Liu RQ, Abbaszade I, Ross H, Burn T, Arner E (2000) The thrombospondin motif of aggrecanase-1 (ADAMTS-4) is critical for aggrecan substrate recognition and cleavage. J Biol Chem **275**:25791–25797

[62] Abbaszade I, Liu RQ, Yang F, Rosenfeld SA, Ross OH, Link JR, Ellis DM, Tortorella MD, Pratta MA, Hollis JM, Wynn R, Duke JL, George HJ, Hillman MC, Jr, Murphy K, Wiswall BH, Copeland RA, Decicco CP, Bruckner R, Nagase H, Itoh Y, Newton RC, Magolda RL, Trzaskos JM, Burn TC (1999) Cloning and characterization of ADAMTS11, an aggrecanase from the ADAMTS family. J Biol Chem **274**:23443–23450

[63] Matthews RT, Gary SC, Zerillo C, Pratta M, Solomon K, Arner EC, Hockfield S (2000) Brain-enriched hyaluronan binding (BEHAB)/brevican cleavage in a glioma cell line is mediated by a disintegrin and metalloproteinase with thrombospondin motifs (ADAMTS) family member. J Biol Chem **275**:22695–22703

[64] Kahn J, Mehraban F, Ingle G, Xin X, Bryant JE, Vehar G, Schoenfeld J, Grimaldi CJ, Peale F, Draksharapu A, Lewin DA, Gerritsen ME (2000) Gene expression profiling in an in vitro model of angiogenesis. Am J Pathol **156**:1887–1900

[65] Tetu B, Brisson J, Lapointe H, Bernard P (1998) Prognostic significance of stromelysin 3, gelatinase A, and urokinase expression in breast cancer. Hum Pathol **29**:979–985

[66] Santavicca M, Noel A, Angliker H, Stoll I, Segain JP, Anglard P, Chretien M, Seidah N, Basset P (1996) Characterization of structural determinants and molecular mechanisms involved in pro-stromelysin-3 activation by 4-aminophenylmercuric acetate and furin-type convertases. Biochem J **315**:953–958

[67] Wolf C, Rouyer N, Lutz Y, et al. (1993) Stromelysin-3 belongs to a subgroup of proteinases expressed in breast carcinoma fibroblastic cells and possibly implicated in tumor progression. Proc Natl Acad Sci USA **90**:1843–1847

[68] Wolf C, Chenard MP, de Grossouvre DP, et al. (1992) Breast-cancer associated stromelysin-3 gene is expressed in basal cell carcinoma and during cutaneous wound healing. J Invest Dermatol **99**:870–872

[69] Muller D, Wolf C, Abecassis J, et al. (1993) Increased stromelysin-3 gene expression is associated with increased local invasiveness in head and neck squamous cell carcinomas. Cancer Res **53**:165–169

[70] Engel G, Heselmeyer K, Auer G, et al. (1994) Correlation between stromelysin-3 mRNA concentration and outcome of human breast cancer. Int J Cancer **58**:830–835

[71] Rouyer N, Wolf C, Chenard MP, et al. (1994) Stromelysin-3 gene expression in human cancer: An overview. Invasion Metastasis **14**:269–275

[72] Pei D, Majumdar G, Weiss SJ (1994) Hydrolytic inactivation of a breast carcinoma cell-derived serpin by human stromelysin-3. J Biol Chem **269**:25849–25855

[73] Finlay TH, Tamir S, Kadner SS, et al. (1993) α1-antitrypsin- and anchorage-independent growth of MCF-7 breast cancer cells. Endocrinology **133**:996–1002

[74] Weiss SJ (1989) Tissue destruction by neutrophils. *N Engl J Med* **320**:365–37

[75] Springer TA (1995) Traffic signals on endothelium for lymphocyte recirculation and leukocyte emigration. Annu Rev Physiol **57**:827–872

[76] Carlos TM, Harlan JM (1994) Leukocyte-endothelial adhesion molecules. Blood **84**:2101–2068

[77] Hynes RO (1992) Integrins: Versatility, modulation, and signaling in cell adhesion. Cell **69**:11–25

[78] O. Hynes (1999) Cell adhesion: Old and new questions. Trends Cell Biol **9**: M33–M37

[79] Ruegg C, Mariotti A (2003) Vascular integrins: Pleiotropic adhesion and signaling molecules in vascular homeostasis and angiogenesis. Cell Mol Life Sci **60**:1135–1157

[80] Stupack DG, Cheresh DA (2002) Get a ligand, get a life: Integrins, signaling and cell survival. J Cell Sci **115**:3729–3738

[81] Juliano RL, Haskill S (1993) Signal transduction from the extracellular matrix. J Cell Biol **120**:577–585

[82] Giancotti FG, Mainiero F (1994) Integrin-mediated adhesion and signaling in tumorigenesis. Biochim Biophys Acta **198**:47–64

[83] Hempstead BL, Birge RB, Fajardo JE, Glassman R, Mahadeo D, Kraemer R, Hanafusa H (1994) Expression of the v-crk oncogene product in PC12 cells results in rapid differentiation by both nerve growth factor- and epidermal growth factor-dependent pathways. Mol Cell Biol **4**:1964–1971

[84] Daemi N, Thomasset N, Lissitzky JC, Dumortier J, Jacquier MF, Pourreyron C, Rousselle P, Chayvialle JA, Remy L (2000) Anti-β4 integrin antibodies enhance migratory and invasive abilities of human colon adenocarcinoma cells and their MMP-2 expression. Int J Cancer **85**:850–856

[85] Khatib AM, Nip J, Fallavollita L, Lehmann M, Jensen G, Brodt P (2001) Regulation of urokinase plasminogen activator/plasmin-mediated invasion of melanoma cells by the integrin vitronectin receptor αvβ3. Int J Cancer **91**:300–308

[86] Lehmann M, Rigot V, Seidah NG, Marvaldi J, Lissitzky JC (1996) Lack of integrin α-chain endoproteolytic cleavage in furin-deficient human colon adenocarcinoma cells LoVo. Biochem J **317**:803–809

[87] Lissitzky JC, Luis J, Munzer JS, Benjannet S, Parat F, Chretien M, Marvaldi J, Seidah NG (2000) Endoproteolytic processing of integrin pro-α subunits involves the redundant function of furin and proprotein convertase (PC) 5A, but not paired basic amino acid converting enzyme (PACE) 4, PC5B or PC7. Biochem J **346**:133–138

[88] Berthet V, Rigot V, Champion S, Secchi J, Fouchier F, Marvaldi J, Luis J (2000) Role of endoproteolytic processing in the adhesive and signaling functions of αvβ5 integrin. J Biol Chem **275**:33308–33333

[89] Mantovani A, Garlanda C, Introna M, Vecchi A (1998) Regulation of endothelial cell function by pro- and anti-inflammatory cytokines. Transplant Proc **30**:4239–4243

[90] Pober JS, Cotran RS (1990) The role of endothelial cells in inflammation. Transplantation **50**:537–544

[91] Balaram SK, Agrawal DK, Edwards JD (1999) Insulin like growth factor-1 activates nuclear factor-kappaB and increases transcription of the intercellular adhesion molecule-1 gene in endothelial cells. Cardiovasc Surg **7**:91–97

[92] Ishizuka T, Takamizawa-Matsumoto M, Suzuki K, Kurita A (1999) Endothelin-1 enhances vascular cell adhesion molecule-1 expression in tumor necrosis factor α-stimulated vascular endothelial cells. Eur J Pharmacol **369**:237–245

[93] Hayasaki Y, Nakajima M, Kitano Y, Iwasaki T, Shimamura T, Iwaki K (1996) ICAM-1 expression on cardiac myocytes and aortic endothelial cells via their specific endothelin receptor subtype 1. Biochem Biophys Res Commun **229**:817–824

[94] McCarron R, Wang L, Stanimirovic DB, Spatz M (1993) Endothelin induction of adhesion molecule expression on human brain microvascular endothelial cells. Neurosci Lett 156:31–34

[95] Morisaki N, Takahashi K, Shiina R, Zenibayashi M, Otabe M, Yoshida S, Saito Y (1994) Platelet-derived growth factor is a potent stimulator of expression of intercellular adhesion molecule-1 in human arterial smooth muscle cells. Biochem Biophys Res Commun 200:612–618

[96] Kim I, Moon SO, Kim SH, Kim HJ, Koh YS, Koh G (2001) VEGF stimulates expression of ICAM-1, VCAM-1 and E-selectin through nuclear factor-kappaB activation in endothelial cells. J Biol Chem 276:7614–7620

[97] Denault JB, Claing A, D'Orleans-Juste P, Sawamura T, Kido T, Masaki T, Leduc R (1995) Processing of proendothelin-1 by human furin convertase. FEBS Lett 362:276–280

[98] Lunn CA, Fan X, Dalie B, Miller K, Zavodny PJ, Narula SK, Lundell D (1997) Purification of ADAM 10 from bovine spleen as a TNFα convertase. FEBS Lett 400:333–335

[99] Izumi Y, Taniuchi Y, Tsuji T, Smith CW, Nakamori S, Fidler IJ, Irimura T (1995) Characterization of human colon carcinoma variant cells selected for sialyl Lex carbohydrate antigen: Liver colonization and adhesion to vascular endothelial cells. Exp Cell Res 216:215–221

[100] Bevilacqua MP, Nelson RM (1993) Selectins: Selectins. J Clin Invest 91:379–387

[101] Lasky LA, Singer MS, Dowbenko D, Imai Y, Henzel WJ, Grimley C, Fennie C, Gillett N, Watson SR, Rosen SD (1992) An endothelial ligand for L-selectin is a novel mucin-like molecule. Cell 69:927–38

[102] Iwai K, Ishikura H, Kaji M, Sugiura H, Ishizu A, Takahashi C, Kato H, Tanabe T, Yoshiki T (1993) Importance of E-selectin (ELAM-1) and sialyl Lewis(a) in the adhesion of pancreatic carcinoma cells to activated endothelium. Int J Cancer 54:972–977

[103] Mannori G, Santoro D, Carter L, Corless C, Nelson RM, Bevilacqua MP (1997) Inhibition of colon carcinoma cell lung colony formation by a soluble form of E-selectin. Am J Pathol 151:233–243

[104] Kuijpers TW, Raleigh M, Kavanagh T, Janssen H, Calafat J, Roos D, Harlan JM (1994) Cytokine-activated endothelial cells internalize E-selectin into a lysosomal compartment of vesiculotubular shape. A tubulin-driven process. J Immunol 152:5060–5069

[105] Balaram SK, Agrawal DK, Allen RT, Kuszynski CA, Edwards JD (1997) Cell adhesion molecules and insulin-like growth factor-1 in vascular disease. J Vasc Surg 25:866–876

[106] Morisaki N, Takahashi K, Shiina R, Zenibayashi M, Otabe M, Yoshida S, Saito Y (1994) Platelet-derived growth factor is a potent stimulator of expression of intercellular adhesion molecule-1 in human arterial smooth muscle cells. Biochem Biophys Res Commun 200:612–618

[107] Christofori G, Semb H (1999) The role of the cell-adhesion molecule E-cadherin as a tumour-suppressor gene. Trends Biochem Sci 24:73–76

[108] Birchmeier W, Behrens J (1994) Cadherin expression in carcinomas: Role in the formation of cell junctions and the prevention of invasiveness. Biochim Biophys Acta 1198:11–26

[109] Bracke ME, Van Roy FM, Mareel MM (1996) The E-cadherin/catenin complex in invasion and metastasis. Curr Top Microbiol Immunol. 213:123–61

[110] Lee SW (1996) H-cadherin, a novel cadherin with growth inhibitory functions and diminished expression in human breast cancer. Nat Med 2:776–782

[111] De Wever O, Westbroek W, Verloes A, Bloemen N, Bracke M, Gespach C, Bruyneel E, Mareel M (2004) Critical role of N-cadherin in myofibroblast invasion and migration in vitro stimulated by colon-cancer-cell-derived TGF-{beta} or wounding. J Cell Sci 117:4691–4703

[112] Shimazui T, Yoshikawa K, Uemura H, Hirao Y, Saga S, Akaza H (2004) The level of cadherin-6 mRNA in peripheral blood is associated with the site of metastasis and with the subsequent occurrence of metastases in renal cell carcinoma. Cancer 101:963–968

[113] Shimazui T, Yoshikawa K, Uemura H, Kawamoto R, Kawai K, Uchida K, Hirao Y, Saga S, Akaza H (2003) Detection of cadherin-6 mRNA by nested RT-PCR as a potential marker for circulating cancer cells in renal cell carcinoma. Int J Oncol. 23:1049–1054

[114] Ozawa M, Kemler R (1990) Correct proteolytic cleavage is required for the cell adhesive function of uvomorulin. J Cell Biol 111:1645–1650

[115] Posthaus H, Dubois CM, Muller E (2003) Novel insights into cadherin processing by subtilisin-like convertases. FEBS Lett **536**:203–208

[116] Tsuji A, Ikoma T, Hashimoto E, Matsuda Y (2002) Development of selectivity of alpha1-antitrypsin variant by mutagenesis in its reactive site loop against proprotein convertase. A crucial role of the P4 arginine in PACE4 inhibition. Protein Eng **15**:123–130

CHAPTER 5

PROPROTEIN CONVERTASES, METALLOPROTEASES AND TUMOR CELL INVASION

DANIEL E. BASSI AND ANDRÉS J.P. KLEIN-SZANTO

Department of Pathology, Fox Chase Cancer Center, Philadelphia, Pennsylvania 19111, USA

Abstract: The destruction of the basement membrane and extra-cellular matrix by various secreted proteinases from malignant and stromal cells is associated with tumor cell invasion and metastasis. The level of expression of these proteinases in tumor cells is associated with advanced-stage tumorigenesis and poor prognosis. Degradation of many extra-cellular matrix components, such as collagen, proteoglycan, fibronectin, vitronectin and laminin facilitate the detachment of tumor cells and their invasiveness. This complex process involves a cascade of proteolytic events in which the primary step likely implicates enzyme activation by the proprotein convertases (PCs). Of the metalloproteinases activated by the PCs of which the expression has been correlated with increased local aggressiveness, metastasis and poor clinical outcome are stromelysin-3 (str-3), membrane-type MMPs (MT-MMPs), the adamalysin metalloproteinases (ADAMs), and the adamalysin metallo-proteinases with thrombospondin motifs (ADAM-TS). All these MMPs possess one or two typical recognition motif for furin-like enzymes and some of them were recently proven experimentally to be cleaved and activated by these enzymes. Thus, the blockade of the activation of these MMPs by PC inhibitors may provide a novel strategy in micrometastasis treatment and prevention

Keywords: Proprotein convertases, MMPs, PDX, furin, PACE4, transgenic mice

1.   INTRODUCTION

Tumor cell invasion is characterized by a complex series of events resulting in the degradation of the extracellular matrix and the subsequent penetration of cancer cells in the underlying or surrounding connective tissue or stroma. Although the acquisition of enhanced cell growth during malignant transformation is of paramount importance to the development and progression of malignancies, tumor cells must disrupt and degrade the extracellular matrix (ECM) in order to invade

A-Majid Khatib (ed.), Regulation of Carcinogenesis, Angiogenesis and Metastasis by the Proprotein Convertases, 89–106.
© 2006 *Springer.*

nearby compartments. Local tissue destruction that parallels tumor cell invasiveness, together with metastasis, or colonization at distant sites from the primary tumor are the major causes of cancer morbidity and mortality. Three basic steps have been identified in the process of invasion [1], attachment to the basement membrane or the stromal matrix; proteolysis of stromal components including collagens and laminins and finally, increased motility to reach blood or lymphatic vessels and gain the general circulation [1]. During the last decade several laboratories, most of which are contributing to this monograph, have described significant roles for proprotein convertases (PCs) during tumor cell development growth and progression. PCs are Ca2+-dependent serine proteases that possess homology to the endoproteases subtilisin (bacteria) and kexin (yeast). The family of PCs is currently comprised of less than a dozen members, known as furin/PACE, PC1/PC3, PC2, PC4, PACE4, PC5/PC6, and PC7/LPC/PC8, SKI/S1P and NARC-1/PCSK9. In this chapter we will focus on those PC-associated processes that relate to tumor cell invasion and metastasis and especially to the PC-metalloproteases-tumor cell invasiveness cascade.

2.   METALLOPROTEINASES AND CANCER

The matrix metalloproteinases (MMPs) are a group of enzymes capable of degrading the different components of the extracellular matrix (ECM). Thus, their expression and regulation (or de-regulation) of their activity is crucial for wound healing, tumor progression and metastasis development [2]. Hence, MMPs catalyze the hydrolytic degradation of collagens, gelatin, aggrecan, versican, laminin and perlecan to cite some of the most representative substrates. All these proteins are major components of the stroma and basement membranes. Degradation of these components leading to the disruption of these structures has serious consequences on tissues biological behavior. In the first place, transformed epithelial cells would be able to invade underlying structures since the functionality of this natural "barrier" in restraining this invasion is diminished [3, 4]. In addition, the products of these stromal proteins degradation are usually biologically active, i.e cell growth, angiogenesis (either stimulating or inhibiting), migration and transformation [5]. Also, after proteolytic degradation the relatively "looser" stroma may release growth factors normally "trapped" within the ECM such as vascular endothelial growth factor (VEGF) and transforming growth factor β (TGF-β) [6].

Under physiological conditions, the activity of these proteins is tightly regulated [7]. In order to avoid hyperactivity, MMPs can be inactivated by autoproteolysis, follow a restricted temporal and spatial pattern of expression (compartmentalization), complex formation with inhibitors as TIMPs (Tissue inhibitors of mealloproteinases) [8] and RECK [9]. On the other hand, these enzymes are synthesized as inactive zymogens, requiring a further proteolytic step to become fully active. Many proteases are responsible for this activation, such as plasmin or other MMPs. Stromelysin-3 and the MT-MMPs (membrane type metalloproteinases) contain the PCs recognition motif and there are many evidences strongly supporting that these enzymes (particularly furin) are responsible for the cleavage and activation of these MMPs.

Tumor cells exert a potent proteolytic activity, degrading the ECM, allowing invasion to the underlying connective tissue. In many cases, the cells in the connective tissue secrete MMP allowing cancer cells to penetrate the stroma. In this context, the identification of MMP as central to the crucial step in the acquisition of an invasive and metastatic phenotype was the first milestone in the study of tumor biology [10]. In addition, MMPs are overexpressed in many types of cancers including breast, pancreas, colon, ovary, skin and in some cases their levels correlate with poor prognosis [11, 12]. In some tumors, MMPs were overexpressed and in an active form before ECM destruction, indicating that their activity is required for the initiation of the invasion process [13].

These evidences indicate that MMPs are de-regulated in cancer, causing a disruption in the normal epithelial-stromal homeostasis.

One particular subgroup is constituted by the Membrane Type MMPs (MT-MMPs). These MMP are associated to the plasma membrane, where they exert their proteolytic activities. One of the most important substrates that are activated by the MT-MMPs is MMP-2. In our prospective, MT-MMPs represent enzymes of paramount importance since all of them contain the PC's recognition site between their pro- and active domain and are putatively activated by PC-like proteases (see below).

3.   CANCER AND PROPROTEIN CONVERTASES

PCs have been associated with cancer since the early 1990's. The first members identified as overexpressed in cancer were the neuroendocrine PCs such as PC1 and PC2. For instance PC2 has been identified by PCR amplification of a human insulinoma cDNA. The putative gene product was reported to have similarity to furin and proposed to function as a pro-hormone convertase [14]. Soon, PC3 was cloned and the analysis of its sequence revealed high homology with PC2 and furin. The authors anticipated that these proteins may play a different role than furin since they were expressed almost exclusively in neuroendocrine tissue and locate in different subcellular compartments [15]. PC2 and PC3 were expressed in pheochromocytomas but not in normal adrenal tissue [16] and their expression correlated with the expression of their substrate, proenkephalin [17], pointing to a direct relationship between these PCs and tumors of the endocrine glands [16].

Before long, these differences in PC expression between normal and tumors or between tumors of different aggressive potentials were observed also in the ubiquitous PCs, furin and PACE4. Furin and PACE4 were highly expressed in lung squamous cell carcinoma and adenocarcinoma. On the other hand, PC1 [3] and PC2 where expressed in small cell lung carcinoma [18]. Overexpression of PCs has also been observed in breast, head and neck and colon tumors. A clear gradation was observed between furin expression, aggressive behavior and VEGF-C expression.

Normal Epithelium   Dysplastic Epithelium   Squamous Cell Carcinoma

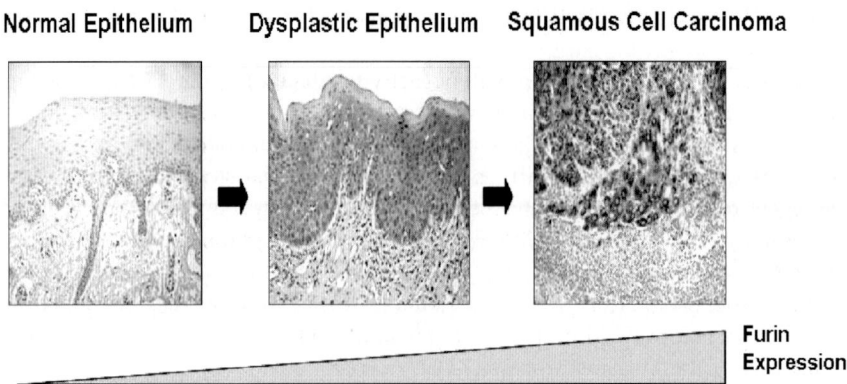

Furin
Expression

Figure 1. Immunohistochemical detection of furin in normal oral epithelium, dysplastic Epithelium and Squamous Cell Carcinoma. Note increased furin expression from normal to dysplatic to carcinoma

Furin expression increased from normal epithelium to invasive SCC in oral epithelia (Figure 1). Moreover furin localized preferentially near or at the invasion front [19]. These correlative studies provided evidence for the association between PCs and cancer. PCs are proteases that cleave proteins at specific sites, modifying the protein substrates leading to alteration of their properties. For instance many proteins have a PC recognition and cleavage site between the pro-domain and the mature protein. Hence, these proteins maturate, usually becoming active, upon PC cleavage. The next step in the development of the ideas that PCs were related to invasion was the discovery of a series of MMP, enzymes responsible for ECM degradation that become activated after PC (specifically furin) cleavage.

The pioneering paper by Pei and Weiss [20] describes the furin-dependent cleavage and activation of Stromelysin-3. This metalloprotease, expressed mainly in stromal cells, had been implicated in tumor progression and degradation of the ECM [21–23]. In contrast to many other MMPs, Stromelysin-3 (MMP-11) contained an insertion of 10 aminoacids GLSA**RNRQKR** which adds a PC cleaving site. This finding was of capital importance since it underscored a direct relationship between furin and PCs and tumor invasion or metastasis. Also, PCs were identified for the first time as possible targets for future therapy.

One year before the previous findings were published, a new MMP gene had been cloned from a cDNA library from human placenta [24]. Analysis of its sequence revealed a unique transmembrane domain; hence this new gene was referred as MT-MMP (for Membrane Type metallopreoteinase). This plasma membrane localization was confirmed by immunohistochemistry. Interestingly, MT-MMP (now renamed as MT1-MMP after the subsequent discovery of five additional members sharing its structural and physiological characteristics) was able to cleave and activate MMP-2, or collagenase A, one of the enzymes responsible for the degradation of collagen IV, a key component of the basement membrane. Surprisingly MT1-MMP had an

insertion cassette similar to Stromelysin-3 generating a PC cleaving site between the pro-domain and the mature form.

These results suggested that a furin-like protease would be responsible for the activation of this new MMP [24]. The crucial role that furin pays in the cleavage and activation of MT1-MMP, was demonstrated shortly after in two independent studies. Sato et al. demonstrated that purified furin was able to process recombinant MT1-MMP in vitro and that this process resulted in stimulation of pro-gelatinase A processing [25]. Pei et al., provided further evidences that MT1-MMP was cleaved at the RRKRY[112] and point mutations within the furin-like cleavage sequence resulted in inhibition of MT1-MMP processing [26]. In agreement with these experiments, treatment of the fibrosarcoma cell line HT-1080 with a synthetic furin inhibitor decreased pro-MT1-MMP processing and pro-MMP-2 processing and activation leading to a reduction in cell invasiveness [27].

More recently Hubbard et al., provided evidence that PACE 4 was overexpressed in murine chemically-induced squamous cell carcinoma of high grade [28]. Moreover, SCC lines transfected with the full-length PACE4 cDNA resulted in the activation of stromelysin-3 and in increased cell invasiveness. These studies broadened the horizons in the field, linking this new family of proteases' expression with the specific processing of matrix-degrading enzymes. Two important substrates, Stromelysin-3 and the family of MT-MMPs were defined. New interactions between metalloproteases and other proteins, some of them PCs' substrates, would come into play increasing the complexity and stressing the role of this novel family in cancer invasion.

4. PROPROTEIN CONVERTASES AND CANCER-RELATED SUBSTRATES

Numerous cancer-related associated proteins are PC substrates. These substrates play significant roles in biological processes that are paramount to tumor promotion and progression, i.e. cell locomotion, adhesion, invasiveness and growth. The most salient substrate groups included in this chapter are those associated with cell to cell and cell to ECM interactions (collagens, integrins, cadherins) as well as those associated with ECM degradation (metalloproteinases). Other PC substrates that also have a significant role during tumor development, namely growth factors and growth factor receptors, will only be reviewed when their activation was shown to impinge on the processes of ECM degradation and tumor cell invasiveness.

Cell locomotion, together with cell adhesiveness and secretion of proteases are crucial elements in the process of tumor cell invasiveness. PCs are involved in all these processes, including cell motility. For instance, furin is capable of processing of some of the collagens associated to membranes [29] as collagen XIII, XVII and XXII. These collagens are important components of hemidesmosomes, structures that participate in cell-cell and cells-stromal adhesion milieu [30]. After

furin cleavage, these proteins ectodomains are shed, influencing negatively on cell adhesion, favoring cell motility and spreading.

PCs were shown to be capable of processing alpha6 [31] and alphaV integrins [32] that play a significant role in cell attachment to stromal components. The relationship between cell attachment and PC expression needs to be further explored to define whether PCs play a significant role in increased attachment to stromal structures and which are the mechanisms whereby increased integrin processing (or other substrates) result in increased attachment.

PCs involvement in the second step, basement membrane proteolysis has been explored in more detail and now it is a field that reached maturity. Overexpression and activation of matrix proteases, i.e., enzymes capable of disrupting the proteins composing the extracellular matrix is a common feature found in tumors. The main targets of these proteases include components of the extracellular matrix itself, such as collagens and laminins as well as other proteolytic enzymes that, after being activated, gain the ability to degrade these and other components of the extracellular matrix. Degradation of these components of the ECM initiates several mechanisms that contribute to an invasive phenotype. First, degradation of the ECM permits epithelial cancer cell access to the underlying connective tissue and eventually to reach blood and lymphatic vessels. On the other hand, many growth factors that are usually "trapped" in the ECM may be released, stimulating transformation and aiding in the cells' invasive behavior. More recently, novel approaches consider that the normal architecture of the basement membrane prevents invasion and, conversely, alteration of this structure is more permissive towards invasion.

The activity of metalloproteinases is an essentially destructive activity that must be highly regulated either at the transcriptional as at the post-translational level. Normal cells do not express high levels of these enzymes. However their expression is higher in wound-healing, tumor cell invasion, and metastasis. In addition, their activity is highly regulated since all of them are synthesized as inactive zymogens that require a further proteolytic step in order to become fully active. PCs activate several matrix metalloproteinases that play a crucial role in ECM degradation, invasion and metastasis. Furthermore, inhibition of PCs activity may constitute a novel therapeutic approach since targeting a single protein or group of protein leads to inhibition of several invasion-related pathways. On the other hand, some of the PCs are capable of degrading some components of the ECM directly, highlighting their role in invasion.

In summary PCs are capable of cleaving a series of substrates resulting in increased stromal degradation and enhanced motility.

4.1   PCs and Metalloproteinases

MT-MMPs, a family of six members, contain a trans-membrane domain that anchors them to the plasma membrane where it is expressed in an active form [24, 33]. All of them contain the cleavage sequence for PCs, namely RRK/RR between their pro and the catalytic domain. The active enzymes catalyze MMP-2 (one of

the major collagen IV degrading enzymes, [34] activation [24, 35–37], collagen I processing [38, 39], direct degradation of ECM components such as the small rich-leucine proteoglycans lumican and decorin [40, 41], adhesion molecules such as integrins [42] and laminin γ2 [41] and molecules involved in cell migration as syndecan [43]. Most of the MT-MMPs studies focused on MT1-MMP, the best known member of the family. Additional studies done with the other members of the MT-MMP family demonstrated that they share with MT1-MMP many biochemical and functional properties.

MT-1 MMP is synthesized as a ~63 kDa zymogen that is subsequently processed and activated at least in some systems by furin or PC5 to a ~57–60 kDa mature protein [44]. Since Sato et al demonstrated that recombinant MT1-MMP was efficiently cleaved in vitro by purified soluble furin [25], this concept has been challenged [42, 45, 46]. Interestingly, furin inhibition either with a synthetic inhibitor (CMK) or antisense technology inhibited Pro-MT1-MMP activation and MMP-2 activity in cardiac and uterine cervical fibroblasts [47, 48] but not in rabbit dermal fibroblasts [48]. These results suggest that furin involvement in MT1-MMP maturation and consequently in the regulation of collagenolytic modulation may depend on the species, tissues and/or cell types.

Inhibition of PCs resulted in inhibition of MT-MMPs processing. Furin transfection to low-grade SCC cell lines from head and neck, resulted in increased levels of MT1-MMP processing. After treatment of these furin overexpressing cells with CMK, furin-mediated MT1-MMP processing was greatly diminished [49]. Furthermore, several SCC head and neck cell lines become 50 to 70% less proficient in MT1-MMP processing after transfection with PDX- a high affinity inhibitor of furin and PC5 [50]. These results were supported with transfection with a second furin specific inhibitor, furin pro-segment (ppfur). Transfection with ppfur resulted in reduction of Pro MT1-MMP cleavage [51]. On the other hand, the extracellular PACE4 was found to cleave MT2-MMP in murine SCC cell lines. When PACE4 expressing SCC and papilloma cell lines were treated with a specific antibody, their ability to process MT2-MMP was diminished [52]. This behavior does not seem to be restricted to SCC since astrocytoma cell lines ability to process MT1-MMP and to invade in vitro and in vivo, was hampered by transfection with PDX [53]. Moreover, only the unprocessed MT1-MMP form was detected when A375 melanoma cell lines were transfected with this inhibitor [54]. Reduced MT-MMPs processing resulted in reduced invasion *in vitro* and *in vivo* [49, 50, 52, 53]. Cells, treated either with CMK, antibodies or transfected with the inhibitors showed a reduction in the ability to degrade and pass through Matrigel-composed by a reconstituted murine extracellular matrix. This behavior was also observed in vivo using the tracheal xenotransplantation assay. This assay permits the evaluation of invasion trough the connective tissue that lies between the ends of the C-shaped tracheal cartilages. In addition, cell proliferation and viability can be assessed. When PDX transfected cells were xenotransplanted into tracheal grafts in immunosuppressed mice, the tumor cells remained viable but their growth and invasive ability were markedly impaired when compared with the vector alone-transfected counterparts.

Other proteases have been suggested to activate MT1-MMP. First, MT1-MMP itself can function as a auto-convertase [46]. However, BB-94 a MT1-MMP inhibitor, was unable to prevent Pro-MT1-MMP cleavage in A375 melanoma cell lines, arguing against the autocatalytic cleavage, at least in this cell line [54]. These results confirm the pivotal role of PCs in metalloproteinase activation and highlight the importance of these proteases as possible targets for novel cancer therapy strategies. Other candidate, plasmin, whose activity is mainly extracellular, was able to cleave Pro-MT1-MMP *in vitro* [55]. However, plasmin, tested at various concentrations had no effect in the generation of active MT1-MMP on control or MT-1MMP overexpressing HT-1080 cell lines [56]. Moreover, incubation of MT1-MMP expressing cells in the presence of plasmin-depleted serum had no effect in the pro-collagenolytic activity of MT1-MMP, indicating that plasmin is not required for MT1-MMP activation [57].

In summary, furin or other PC convertases are likely to be the physiological activators of this group of metalloproteinases. This action seems to be constitutive although a model of the regulation of furin expression through TGF-β has been proposed suggesting that the activation of MT-MMPs might be dependent, in some systems, on the levels of serum TGF-β [58].

Recently, it has been shown in co-transfection systems, that furin cleaves MMP-2 directly in the trans-Golgi network, and the product of this proteolysis is an inactive metalloprotease [59]. This finding provides a novel view at the role of furin (and PCs) in invasion, since convertases are proposed as preventing rather that supporting invasion. Although further studies need to be performed, it can be speculated that furin contributes to directing the process of invasion by activating substrates in the compartment where they are needed (i.e., the extracellular) while preventing their putative devastating action in the intracellular compartments. In addition, it is probable that MMP-2 and furin are expressed in different cell types, i.e stromal and epithelial, and that the interaction between them, one secreting the pro-MMP, and the other cell type providing the machinery to activate this Pro-MMP, results in enhancement of cell invasion.

4.2   PC and Other Sustrates Related to Tumor Cell Invasiveness

4.2.1   IGF-1R

IGF-1R is synthesized as a 200 kb precursor that, after furin cleavage, originates the α and β chains of the mature receptor [60] [61]. Cleavage of IGF-1R precursor is absolutely necessary for intracellular signaling (tyrosine kinase activity) and ligand (IGF-1) binding [62]. In addition, inhibition of furin activity results in decreased levels of receptor maturation, leading to resistance to IGF-1 mediated proliferation [49, 61] suggesting that furin-mediated cleavage is essential for IGF-1R functionality. Other PCs are capable of cleaving Pro-IGF-1R, albeit with much less efficiency as shown in LoVo cells, which do not express functional furin. In these cells, IGF-1R processing was greatly reduced but not totally abolished, pointing to the action of other (s) PC [62].

Binding of IGF-1R ligand, IGF-1, leads to increased proliferation as well as pro-survival signals. In addition, lung carcinoma cells over expressing this receptor showed enhanced invasion through Matrigel indicating that IGF-1R favored degradation of some components of the basement membrane, probably through induction of MMPs. In this context, Long et al demonstrated that over-expression of IGF-1R in M-27 lung cancer cell lines, which express low levels of the receptor, resulted in increased MMP-2, levels. Conversely, ablation of IGF-1R expression using antisense technology, lowered MMP-2 levels resulting in impaired invasive ability [63]. Recently, a dual regulation of MMP-2 levels through the IGF-1/IGF-1R signal transduction system has been elucidated [64]. Two major pathways contribute to MMP-2 regulation: the PI 3 kinase/Akt/mTOR and Ras/Raf/MEK pathaways. When H59 lung carcinoma cells were stimulated with IGF-1concentrations (10 ng/ml) that favor cell proliferation and motility, activation of PI 3-kinase/Akt/mTOR resulted in increased MMP-2 mRNA and protein synthesis. The other pathway, the Ras/Raf/MEK axis, is also activated but in a transient and weak manner, partly due to Akt-mediated inactivation of Raf-1. Interestingly, at higher IGF-1 doses (100 ng/ml), the latter pathway is preponderant, leading mainly to MMP-2 suppression in these cells. These two contrasting activities exerted by different IGF-1 concentrations emphasizes IGF-1's role in fine-tuning the invasive ability of cancer cells.

Different cell lines may respond differently to IGF-1. In MCF-7 breast cancer cells, high doses of IGF-1 (100 ng/ml) activated the PI 3Kinase pathway, resulting in Raf phosphorylation at Ser 259, inhibiting its activity [65]. Although free (not bound to IGF binding proteins), IGF-1 concentrations in tumors may be difficult to assess. These results point to a critical role of this grow factor in the acquisition of the invasive phenotype. In a similar way, the number of active receptor in the plasma membrane should play a decisive role in MMP-2 induction. Furin, and to a lesser extent PC5, cleaves and activates IGF-1R [61]. Inhibition of furin activity led to reduced levels of active receptor, less proliferation, and resistance to IGF-1. Ideally, total ablation of IGF-1R functionality implies a complete irresponsiveness to its ligand, IGF-1, regardless of IGF-1 concentrations in tumor tissue. This will result in decreased PI3 kinase activation and subsequently decreased MMP-2 levels, thus inhibiting tumor cell invasiveness.

4.2.2   TGF-β

TGF-β remains one of the most controversial molecules in the field of cancer invasion and metastases. Numerous papers have been published supporting its role in increasing motility, invasion and metastasis. Equally number of works challenged this concept arguing that the opposite actually happens; decreased tumorigenesis and suppression of metastasis. To make matters worse, it is possible to find these contradicting results in similar tumor cell lines.

To cite two of the more recent and prominent works in the field, Muraoka-Cook et al. using a transgenic model where TGF-β is expressed conditionally in the mammary gland, demonstrated that expression of this growth factor resulted in >10 fold metastasis to lung whereas primary cells proliferation or tumor size was unaffected by the expression of the transgene [66]. Conversely, abolition of the TGF-β pathway, by disrupting TGF-β receptor II in the mouse mammary gland, resulted in increased lung metastasis in a conditional knockout model [67].

It is probable that these conflicting ideas may be reconciled by taking into account a considerable number of evidences pointing to a dual role of TGF-β; suppressor of tumorigenesis in early stages of tumor progression, (pre-malignant or well-differentiated cells). On the other hand, it is an enhancer of tumor development, growth and metastasis in more aggressive or later stages [68–70].

Since 1995 when Dubois et al. provided convincing evidence that furin was responsible of the first step in TGF-β maturation in vitro and in vivo [71, 72], the field has been steadily growing. TGF-β induces the expression of metalloproteinases and integrins playing an important role in the acquisition of the malignant phenotype. It has been shown that TGF-β induces alpha 3 integrins in hepatocellular carcinoma leading to a more invasive behavior [73]. The fact that active TGF-β is only secreted in invasive hepatocellular carcinoma cell lines but not in non-invasive or normal liver cells [73] and that furin and TGF-β co-localized in regenerating rat liver after partial hepatectomy where TGF-β is required in an active form [74] suggest that the TGF-β-driven increase in cell growth and invasion may be furin-dependent.

Furin mediates increased MMP-2 activation, and hence invasion, through a MT-MMP independent, TGF-β dependent pathway [75]. This findings lead to the proposition of a new axis that considers TGF-β as a new mediator between MMPs and furin.

4.3   Miscellaneous Substrates

Many other proteins containing the furin-cleavage recognition sequence and at the same time putatively associated with invasion and metastasis have been identified. The ADAM (a disintegrin and metalloproteinase) and ADAMTS (a disintegrin and metalloproteinase domain, with thrombospondin type-1 modules) families are proteases associated with cell-surface proteolysis and adhesion. The inflammatory cytokine TACE or tumor necrosis factor-alpha converting enzyme was the first member of this family to be discovered. Interestingly it is processed and activated by furin and other PCs such as PACE4, PC1/PC2 and PC5 [76, 77]. After this first discovery, many other ADAMs were identified. Some of them are furin substrates and are overexpressed in cancer [78–80]. For instance, ADAM9, which may be putatively activated by furin, is expressed in non-small cell lung cancer and its expression correlates with brain metastasis. The mechanism of action involves modulation of adhesion molecules such as $\alpha_3\beta_1$ integrins [81].

Processing by furin is a pre-requisite for ADAM 10 [82] and 12 activity [83] and, probably for ADAM 15 [84]. These new findings linking furin to novel mechanisms of adhesion and proteolytic degradation open a new field that is currently actively explored. Recently, it has been demonstrated that furin and possibly other PCs are responsible for shedding membrane-anchored proteins, such as membrane-associated collagens, modifying the adhesion properties of tumor cells. Because of its significance for cell invasion and its relationship to furin more studies will be required in the future. Whether furin cleavage is necessary in order to alter the functionality of these collagens and whether or not this cleavage aids in tumor cell invasiveness awaits further clarification.. In this context, it has been shown that furin-dependent cleavage of collagen XXIII results in its shedding and multi-merization, a feature already observed in prostate cancer [85]. Others have argued that the furin "sheddase" activity may be ADAM-dependent, at least in the case of collagen XVII [86].

4.4   Crosstalk between PC–Activated Pathways

MT1-MMP can be induced in response to insulin like growth factor I (IGF-1) stimu-lation in a pathway that involves PI 3 kinase/Akt signaling [87]. In addition, furin cleaves pro-IGF-1R into α and β chains [61], a prerequisite for the receptor to be

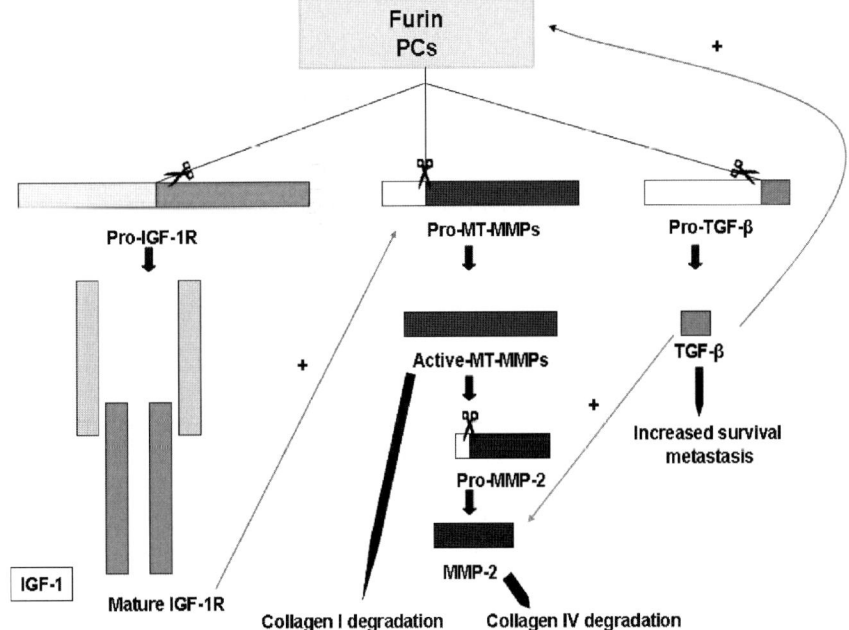

Figure 2. Crosstalk between PC–Activated Pathways

functional. In this context, furin also contributes to MT1-MMP activity through increased MT1-MMP synthesis, providing additional reinforcement of the basic axis PC/MT-MMPs/MMP-2 activation cascade. Recently, it has been shown that IGF-1/PI 3 kinase pathway was involved in the induction of MT1-MMP in vascular smooth muscle cells [88]. Interestingly, IGF-1 mediated induction of MT1-MMP was inhibited by treatment with CMK, a general PC inhibitor in this system, suggesting that inhibition of PC may derive in diminishing MT1-MMP activity through lesser levels of functional IGF-1R and, thus reduced degradation of ECM components as collagen I and IV. Moreover, TGF-β induces the MMP-2 expression linking this growth factor to the furin/MT-MMPs/MMP-2 axis. Interestingly, TGF-β seems to regulate furin expression, positioning it upstream all these pathways related to invasion and metastasis. These relationships are summarized in Figure 2.

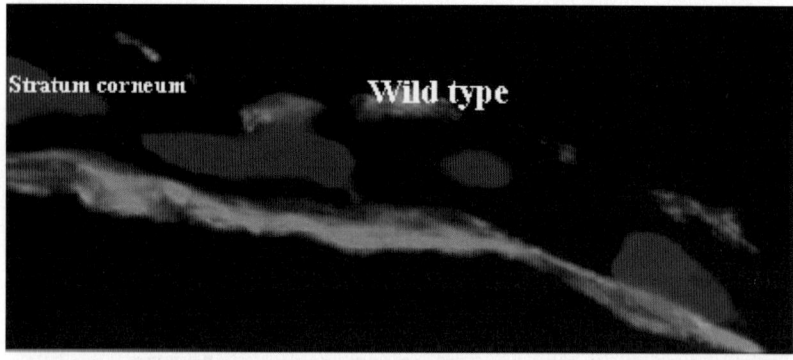

Figure 3. Collagen IV staining. Frozen sections from skins form wild type or PACE4 transgenic mice were stained with a monoclonal antibody against collagen IV. Collagen IV was visualized by immunofluorescense (shown in green). Nuclear stain was performed with Hoechst H 33342 (in blue). The collagen IV component of the basement membrane is markedly thinner and more disrupted (arrows) than the wild type that exhibits a continuous collagen IV layer. The fluorescence intensity was normalized to the staining of the stratum corneum, considered the internal control

5. ANIMAL MODELS

The use of animal models is emerging as a scientific necessity to confirm the role of PC's in mammalian tissues. The classical approach is the production of either transgenic mice overexpressing PCs or the ablation of the gene in knock-out models.

Knocking down furin expression is lethal at early embryonic stages due to failure in heart development and other hemodynamic defects [89]. This study provided strong evidences favoring the non-redundancy in PCs activity, since this lethal phenotype could not be overcome by the other PCs that were expressed at normal levels. On the other hand, it was clear that, in order to study the physiological substrates of furin, an inducible knock-out model was necessary. Recently, an interferon inducible furin knock-out mouse has been developed [90]. The authors centered their study in the effects of furin ablation in liver, because of its numerous metabolic activities. Abrogation of furin activity did not result in any major histological changes but processing of some substrates as alpha 5 integrin or alpha1-microglobulin was diminished, although not completely abolished. These data pointed to some redundance in PCs physiological activities and anticipated that therapeutic approaches inhibiting furin may stop the progression of diseases without exhibiting major secondary toxic effects. This model will be extremely valuable to evaluate whether furin deficiency protects from liver tumor progression, invasion or metastasis using chemical-induced protocols of murine hepatocellular carcinogenesis.

In our laboratory we produced a transgenic model overexpressing PACE4 in the basal layer of the epidermis. Although no major phenotypic changes have been observed in these PACE4-overexpressing animals, histological examination of the skins revealed disruption of the collagen IV component of the basement membrane (Figure 3) which coincided with increased processing of MT1 and MT2-MMP. These changes resulted in the development of higher number of aggressive tumors and metastasis (Bassi, DE et al, unpublished results).

REFERENCES

[1] Stetler-Stevenson WG, Yu AE (2001) Proteases in invasion: Matrix metalloproteinases. Semin Cancer Biol **11**:143–152
[2] Matrisian LM (1990) Metalloproteinases and their inhibitors in matrix remodeling. Trends Genet **6**:121–125
[3] Chang C, Werb Z (2001) The many faces of metalloproteases: Cell growth, invasion, angiogenesis and metastasis. Trends Cell Biol **11**:S37–43
[4] Mueller MM, Fusenig NE (2004) Friends or foes - bipolar effects of the tumour stroma in cancer. Nat Rev Cancer **4**:839–849
[5] Schenk S, Quaranta V (2003) Tales from the crypt[ic] sites of the extracellular matrix. Trends Cell Biol **13**:366–375
[6] Mott JD, Werb Z (2004) Regulation of matrix biology by matrix metalloproteinases. Curr Opin Cell Biol **16**:558–564
[7] Chakraborti S, Mandal M, Das S, Mandal A, Chakraborti T (2003) Regulation of matrix metallo-proteinases: An overview. Mol Cell Biochem **253**:269–285

[8] Baker AH, Edwards DR, Murphy G (2002) Metalloproteinase inhibitors: Biological actions and therapeutic opportunities. J Cell Sci **115**:3719–3727

[9] Noda M, Oh J, Takahashi R, Kondo S, Kitayama H, Takahashi C (2003) RECK: A novel suppressor of malignancy linking oncogenic signaling to extracellular matrix remodeling. Cancer Metastasis Rev **22**:167–175

[10] Liotta LA, Tryggvason K, Garbisa S, Hart I, Foltz CM, Shafie S (1980) Metastatic potential correlates with enzymatic degradation of basement membrane collagen. Nature **284**:67–68

[11] Basset P, Okada A, Chenard MP, Kannan R, Stoll I, Anglard P, Bellocq JP, Rio MC (1997) Matrix metalloproteinases as stromal effectors of human carcinoma progression: therapeutic implications. Matrix Biol **15**:535–541

[12] Westermarck J, Kahari VM (1999) Regulation of matrix metalloproteinase expression in tumor invasion. Faseb J **13**:781–792

[13] Nelson AR, Fingleton B, Rothenberg ML, Matrisian LM (2000) Matrix metalloproteinases: Biologic activity and clinical implications. J Clin Oncol **18**:1135–1149

[14] Smeekens SP, Steiner DF (1990) Identification of a human insulinoma cDNA encoding a novel mammalian protein structurally related to the yeast dibasic processing protease Kex2. J Biol Chem **265**:2997–3000

[15] Smeekens SP, Avruch AS, LaMendola J, Chan SJ, Steiner DF (1991) Identification of a cDNA encoding a second putative prohormone convertase related to PC2 in AtT20 cells and islets of Langerhans. Proc Natl Acad Sci USA **88**:340–344

[16] Konoshita T, Gasc JM, Villard E, Takeda R, Seidah NG, Corvol P, Pinet F (1994) Expression of PC2 and PC1/PC3 in human pheochromocytomas. Mol Cell Endocrinol **99**:307–314

[17] Breslin MB, Lindberg I, Benjannet S, Mathis JP, Lazure C, Seidah NG (1993) Differential processing of proenkephalin by prohormone convertases 1(3) and 2 and furin. J Biol Chem **268**:27084–27093

[18] Mbikay M, Sirois F, Yao J, Seidah NG, Chretien M (1997) Comparative analysis of expression of the proprotein convertases furin, PACE4, PC1 and PC2 in human lung tumours. Br J Cancer **75**:1509–1514

[19] Lopez de Cicco R, Watson JC, Bassi DE, Litwin S, Klein-Szanto AJ (2004) Simultaneous expression of furin and vascular endothelial growth factor in human oral tongue squamous cell carcinoma progression. Clin Cancer Res **10**:4480–4488

[20] Pei D, Weiss SJ (1995) Furin-dependent intracellular activation of the human stromelysin-3 zymogen. Nature **375**:244–247

[21] Basset P, Bellocq JP, Wolf C, Stoll I, Hutin P, Limacher JM, Podhajcer OL, Chenard MP, Rio MC, Chambon P (1990) A novel metalloproteinase gene specifically expressed in stromal cells of breast carcinomas. Nature **348**:699–704

[22] Wolf C, Chenard MP, Durand de Grossouvre P, Bellocq JP, Chambon P, Basset P (1992) Breast-cancer-associated stromelysin-3 gene is expressed in basal cell carcinoma and during cutaneous wound healing. J Invest Dermatol **99**:870–872

[23] Muller D, Wolf C, Abecassis J, Millon R, Engelmann A, Bronner G, Rouyer N, Rio MC, Eber M, Methlin G, et al. (1993) Increased stromelysin 3 gene expression is associated with increased local invasiveness in head and neck squamous cell carcinomas. Cancer Res **53**:165–169

[24] Sato H, Takino T, Okada Y, Cao J, Shinagawa A, Yamamoto E, Seiki M (1994) A matrix metalloproteinase expressed on the surface of invasive tumour cells. Nature **370**:61–65

[25] Sato H, Kinoshita T, Takino T, Nakayama K, Seiki M (1996) Activation of a recombinant membrane type 1-matrix metalloproteinase (MT1-MMP) by furin and its interaction with tissue inhibitor of metalloproteinases (TIMP)-2. FEBS Lett **393**:101–104

[26] Pei D, Weiss SJ (1996) Transmembrane-deletion mutants of the membrane-type matrix metalloproteinase-1 process progelatinase A and express intrinsic matrix-degrading activity. J Biol Chem **271**:9135–9140

[27] Maquoi E, Noel A, Frankenne F, Angliker H, Murphy G, Foidart JM (1998) Inhibition of matrix metalloproteinase 2 maturation and HT1080 invasiveness by a synthetic furin inhibitor. FEBS Lett **424**:262–466

[28] Hubbard FC, Goodrow TL, Liu SC, Brilliant MH, Basset P, Mains RE, Klein-Szanto AJ (1997) Expression of PACE4 in chemically induced carcinomas is associated with spindle cell tumor conversion and increased invasive ability. Cancer Res 57:5226–5231

[29] Pihlajaniemi T, Rehn M (1995) Two new collagen subgroups: Membrane-associated collagens and types XV and XVII. Prog Nucleic Acid Res Mol Biol 50:225–262

[30] Hopkinson SB, Baker SE, Jones JC (1995) Molecular genetic studies of a human epidermal autoantigen (the 180-kD bullous pemphigoid antigen/BP180): Identification of functionally important sequences within the BP180 molecule and evidence for an interaction between BP180 and alpha 6 integrin. J Cell Biol 130:117–125

[31] Bergeron E, Basak A, Decroly E, Seidah NG (2003) Processing of alpha4 integrin by the proprotein convertases: Histidine at position P6 regulates cleavage. Biochem J 373:475–384

[32] Mayer G, Boileau G, Bendayan M (2003) Furin interacts with proMT1-MMP and integrin alphaV at specialized domains of renal cell plasma membrane. J Cell Sci 116:1763–1773

[33] Polette M, Birembaut P (1998) Membrane-type metalloproteinases in tumor invasion. Int J Biochem Cell Biol 30:1195–1202

[34] Collier IE, Wilhelm SM, Eisen AZ, Marmer BL, Grant GA, Seltzer JL, Kronberger A, He CS, Bauer EA, Goldberg GI (1988) H-ras oncogene-transformed human bronchial epithelial cells (TBE-1) secrete a single metalloprotease capable of degrading basement membrane collagen. J Biol Chem 263:6579–6587

[35] Takino T, Sato H, Shinagawa A, Seiki M (1995) Identification of the second membrane-type matrix metalloproteinase (MT-MMP-2) gene from a human placenta cDNA library. MT-MMPs form a unique membrane-type subclass in the MMP family. J Biol Chem 270:23013–23020

[36] Cao J, Sato H, Takino T, Seiki M (1995) The C-terminal region of membrane type matrix metalloproteinase is a functional transmembrane domain required for pro-gelatinase A activation. J Biol Chem 270:801–805

[37] Butler GS, Will H, Atkinson SJ, Murphy G (1997) Membrane-type-2 matrix metalloproteinase can initiate the processing of progelatinase A and is regulated by the tissue inhibitors of metalloproteinases. Eur J Biochem 244:653–657

[38] Ohuchi E, Imai K, Fujii Y, Sato H, Seiki M, Okada Y (1997) Membrane type 1 matrix metalloproteinase digests interstitial collagens and other extracellular matrix macromolecules. J Biol Chem 272:2446–2451

[39] Hotary K, Allen E, Punturieri A, Yana I, Weiss SJ (2000) Regulation of cell invasion and morphogenesis in a three-dimensional type I collagen matrix by membrane-type matrix metalloproteinases 1, 2, and 3. J Cell Biol 149:1309–1323

[40] Li Y, Aoki T, Mori Y, Ahmad M, Miyamori H, Takino T, Sato H (2004) Cleavage of lumican by membrane-type matrix metalloproteinase-1 abrogates this proteoglycan-mediated suppression of tumor cell colony formation in soft agar. Cancer Res 64:7058–7064

[41] Koshikawa N, Giannelli G, Cirulli V, Miyazaki K, Quaranta V (2000) Role of cell surface metalloprotease MT1-MMP in epithelial cell migration over laminin-5. J Cell Biol 148:615–624

[42] Ratnikov BI, Rozanov DV, Postnova TI, Baciu PG, Zhang H, DiScipio RG, Chestukhina GG, Smith JW, Deryugina EI, Strongin AY (2002) An alternative processing of integrin alpha(v) subunit in tumor cells by membrane type-1 matrix metalloproteinase. J Biol Chem 277:7377–7385

[43] Endo K, Takino T, Miyamori H, Kinsen H, Yoshizaki T, Furukawa M, Sato H (2003) Cleavage of syndecan-1 by membrane type matrix metalloproteinase-1 stimulates cell migration. J Biol Chem 278:40764–40770

[44] Nagase H, Woessner JF, Jr (1999) Matrix metalloproteinases. J Biol Chem 274:21491–21494

[45] Rozanov DV, Deryugina EI, Ratnikov BI, Monosov EZ, Marchenko GN, Quigley JP, Strongin AY (2001) Mutation analysis of membrane type-1 matrix metalloproteinase (MT1-MMP). The role of the cytoplasmic tail Cys(574), the active site Glu(240), and furin cleavage motifs in oligomerization, processing, and self-proteolysis of MT1-MMP expressed in breast carcinoma cells. J Biol Chem 276:25705–25714

[46] Rozanov DV, Strongin AY (2003) Membrane type-1 matrix metalloproteinase functions as a proprotein self-convertase. Expression of the latent zymogen in Pichia pastoris, autolytic activation, and the peptide sequence of the cleavage forms. J Biol Chem 278:8257–8260

[47] Guo C, Jiang J, Elliott JM, Piacentini L (2005) Paradigmatic identification of MMP-2 and MT1-MMP activation systems in cardiac fibroblasts cultured as a monolayer. J Cell Biochem 94:446–459

[48] Sato T, Kondo T, Fujisawa T, Seiki M, Ito A (1999) Furin-independent pathway of membrane type 1-matrix metalloproteinase activation in rabbit dermal fibroblasts. J Biol Chem 274:37280–37284

[49] Bassi DE, Mahloogi H, Lopez De Cicco R, Klein-Szanto A (2003) Increased furin activity enhances the malignant phenotype of human head and neck cancer cells. Am J Pathol 162:439–447

[50] Bassi DE, Lopez De Cicco R, Mahloogi H, Zucker S, Thomas G, Klein-Szanto AJ (2001) Furin inhibition results in absent or decreased invasiveness and tumorigenicity of human cancer cells. Proc Natl Acad Sci USA 98:10326–10331

[51] Lopez de Cicco R, Bassi DE, Zucker S, Seidah NG, Klein-Szanto AJ (2005) Human carcinoma cell growth and invasiveness is impaired by the propeptide of the ubiquitous proprotein convertase furin. Cancer Res 65:4162–4171

[52] Mahloogi H, Bassi DE, Klein-Szanto AJ (2002) Malignant conversion of non-tumorigenic murine skin keratinocytes overexpressing PACE4. Carcinogenesis 23:565–572

[53] Mercapide J, Lopez De Cicco R, Bassi DE, Castresana JS, Thomas G, Klein-Szanto AJ (2002) Inhibition of furin-mediated processing results in suppression of astrocytoma cell growth and invasiveness. Clin Cancer Res 8:1740–1746

[54] Mazzone M, Baldassarre M, Beznoussenko G, Giacchetti G, Cao J, Zucker S, Luini A, Buccione R (2004) Intracellular processing and activation of membrane type 1 matrix metalloprotease depends on its partitioning into lipid domains. J Cell Sci 117:6275–87

[55] Okumura Y, Sato H, Seiki M, Kido H (1997) Proteolytic activation of the precursor of membrane type 1 matrix metalloproteinase by human plasmin. A possible cell surface activator. FEBS Lett 402:181–184

[56] Monea S, Lehti K, Keski-Oja J, Mignatti P (2002) Plasmin activates pro-matrix metalloproteinase-2 with a membrane-type 1 matrix metalloproteinase-dependent mechanism. J Cell Physiol 192:160–170

[57] Chun TH, Sabeh F, Ota I, Murphy H, McDonagh KT, Holmbeck K, Birkedal-Hansen H, Allen ED, Weiss SJ (2004) MT1-MMP-dependent neovessel formation within the confines of the three-dimensional extracellular matrix. J Cell Biol 167:757–767

[58] Blanchette F, Day R, Dong W, Laprise MH, Dubois CM (1997) TGFbeta1 regulates gene expression of its own converting enzyme furin. J Clin Invest 99:1974–1983

[59] Cao J, Rehemtulla A, Pavlaki M, Kozarekar P, Chiarelli C (2005) Furin directly cleaves proMMP-2 in the trans-Golgi network resulting in a nonfunctioning proteinase. J Biol Chem 280:10974–10980

[60] Ullrich A, Gray A, Tam AW, Yang-Feng T, Tsubokawa M, Collins C, Henzel W, Le Bon T, Kathuria S, Chen E, et al. (1986) Insulin-like growth factor I receptor primary structure: comparison with insulin receptor suggests structural determinants that define functional specificity. Embo J 5:2503–2512

[61] Khatib AM, Siegfried G, Prat A, Luis J, Chretien M, Metrakos P, Seidah NG (2001) Inhibition of proprotein convertases is associated with loss of growth and tumorigenicity of HT-29 human colon carcinoma cells: Importance of insulin-like growth factor-1 (IGF-1) receptor processing in IGF-1-mediated functions. J Biol Chem 276:30686–30693

[62] Lehmann M, Andre F, Bellan C, Remacle-Bonnet M, Garrouste F, Parat F, Lissitsky JC, Marvaldi J, Pommier G (1998) Deficient processing and activity of type I insulin-like growth factor receptor in the furin-deficient LoVo-C5 cells. Endocrinology 139:3763–3771

[63] Long L, Navab R, Brodt P (1998) Regulation of the Mr 72,000 type IV collagenase by the type I insulin-like growth factor receptor. Cancer Res 58:3243–3247

[64] Zhang D, Bar-Eli M, Meloche S, Brodt P (2004) Dual regulation of MMP-2 expression by the type 1 insulin-like growth factor receptor: the phosphatidylinositol 3-kinase/Akt and Raf/ERK pathways transmit opposing signals. J Biol Chem 279:19683–19690

[65] Moelling K, Schad K, Bosse M, Zimmermann S, Schweneker M (2002) Regulation of Raf-Akt Cross-talk. J Biol Chem 277:31099–31106

[66] Muraoka-Cook RS, Kurokawa H, Koh Y, Forbes JT, Roebuck LR, Barcellos-Hoff MH, Moody SE, Chodosh LA, Arteaga CL (2004) Conditional overexpression of active transforming growth factor beta1 in vivo accelerates metastases of transgenic mammary tumors. Cancer Res **64**:9002–9011

[67] Forrester E, Chytil A, Bierie B, Aakre M, Gorska AE, Sharif-Afshar AR, Muller WJ, Moses HL (2005) Effect of conditional knockout of the type II TGF-beta receptor gene in mammary epithelia on mammary gland development and polyomavirus middle T antigen induced tumor formation and metastasis. Cancer Res **65**:2296–2302

[68] Tian F, DaCosta Byfield S, Parks WT, Yoo S, Felici A, Tang B, Piek E, Wakefield LM, Roberts AB (2003) Reduction in Smad2/3 signaling enhances tumorigenesis but suppresses metastasis of breast cancer cell lines. Cancer Res **63**:8284–8292

[69] Weeks BH, He W, Olson KL, Wang XJ (2001) Inducible expression of transforming growth factor beta1 in papillomas causes rapid metastasis. Cancer Res **61**:7435–7443

[70] Muraoka-Cook RS, Dumont N, Arteaga CL (2005) Dual role of transforming growth factor beta in mammary tumorigenesis and metastatic progression. Clin Cancer Res **11**:937s–943s

[71] Dubois CM, Laprise MH, Blanchette F, Gentry LE, Leduc R (1995) Processing of transforming growth factor beta 1 precursor by human furin convertase. J Biol Chem **270**:10618–10624

[72] Dubois CM, Blanchette F, Laprise MH, Leduc R, Grondin F, Seidah NG (2001) Evidence that furin is an authentic transforming growth factor-beta1-converting enzyme. Am J Pathol **158**:305–316

[73] Giannelli G, Fransvea E, Marinosci F, Bergamini C, Colucci S, Schiraldi O, Antonaci S (2002) Transforming growth factor-beta1 triggers hepatocellular carcinoma invasiveness via alpha3beta1 integrin. Am J Pathol **161**:183–193

[74] Hoshino H, Konda Y, Takeuchi T (1997) Co-expression of the proprotein-processing endoprotease furin and its substrate transforming growth factor beta1 and the differentiation of rat hepatocytes. FEBS Lett **419**:9–12

[75] McMahon S, Laprise MH, Dubois CM (2003) Alternative pathway for the role of furin in tumor cell invasion process. Enhanced MMP-2 levels through bioactive TGFbeta. Exp Cell Res **291**:326–339

[76] Peiretti F, Canault M, Deprez-Beauclair P, Berthet V, Bonardo B, Juhan-Vague I, Nalbone G (2003) Intracellular maturation and transport of tumor necrosis factor alpha converting enzyme. Exp Cell Res **285**:278–285

[77] Srour N, Lebel A, McMahon S, Fournier I, Fugere M, Day R, Dubois CM (2003) TACE/ADAM-17 maturation and activation of sheddase activity require proprotein convertase activity. FEBS Lett **554**:275–283

[78] McCulloch DR, Harvey M, Herington AC (2000) The expression of the ADAMs proteases in prostate cancer cell lines and their regulation by dihydrotestosterone. Mol Cell Endocrinol **167**:11–21

[79] O'Shea C, McKie N, Buggy Y, Duggan C, Hill AD, McDermott E, O'Higgins N, Duffy MJ (2003) Expression of ADAM-9 mRNA and protein in human breast cancer. Int J Cancer **105**:754–761

[80] Schutz A, Hartig W, Wobus M, Grosche J, Wittekind C, Aust G (2005) Expression of ADAM15 in lung carcinomas. Virchows Arch **446**:421–429

[81] Shintani Y, Higashiyama S, Ohta M, Hirabayashi H, Yamamoto S, Yoshimasu T, Matsuda H, Matsuura N (2004) Overexpression of ADAM9 in non-small cell lung cancer correlates with brain metastasis. Cancer Res **64**:4190–4196

[82] Anders A, Gilbert S, Garten W, Postina R, Fahrenholz F (2001) Regulation of the alpha-secretase ADAM10 by its prodomain and proprotein convertases. Faseb J **15**:1837–1839

[83] Loechel F, Gilpin BJ, Engvall E, Albrechtsen R, Wewer UM (1998) Human ADAM 12 (meltrin alpha) is an active metalloprotease. J Biol Chem **273**:16993–16997

[84] Lum L, Reid MS, Blobel CP (1998) Intracellular maturation of the mouse metalloprotease disintegrin MDC15. J Biol Chem **273**:26236–26247

[85] Banyard J, Bao L, Zetter BR (2003) Type XXIII collagen, a new transmembrane collagen identified in metastatic tumor cells. J Biol Chem **278**:20989–20994

[86] Franzke CW, Tasanen K, Borradori L, Huotari V, Bruckner-Tuderman L (2004) Shedding of collagen XVII/BP180: Structural motifs influence cleavage from cell surface. J Biol Chem **279**:24521–24529

[87] Zhang D, Brodt P (2003) Type 1 insulin-like growth factor regulates MT1-MMP synthesis and
      tumor invasion via PI 3-kinase/Akt signaling. Oncogene **22**:974–982
[88] Stawowy P, Kallisch H, Kilimnik A, Margeta C, Seidah NG, Chretien M, Fleck E, Graf K (2004)
      Proprotein convertases regulate insulin-like growth factor 1-induced membrane-type 1 matrix
      metalloproteinase in VSMCs via endoproteolytic activation of the insulin-like growth factor-1
      receptor. Biochem Biophys Res Commun **321**:531–538
[89] Roebroek AJ, Umans L, Pauli IG, Robertson EJ, van Leuven F, Van de Ven WJ, Constam DB
      (1998) Failure of ventral closure and axial rotation in embryos lacking the proprotein convertase
      Furin. Development **125**:4863–4876
[90] Roebroek AJ, Taylor NA, Louagie E, Pauli I, Smeijers L, Snellinx A, Lauwers A, Van de Ven WJ,
      Hartmann D, Creemers JW (2004) Limited redundancy of the proprotein convertase furin in mouse
      liver. J Biol Chem **279**:53442–53450

CHAPTER 6

MODULATION OF INTEGRIN FUNCTION BY ENDOPROTEOLYTIC PROCESSING: ROLE IN TUMOUR PROGRESSION

RIGOT V. AND LUIS J.

CNRS FRE2737, Faculté de Pharmacie, 27 Bd J. Moulin, 13 385 Marseille Cedex 5, France

Abstract: Integrins are heterodimeric proteins composed by the non-covalent association of α and β subunits. Some integrin α chains undergo a post-translational cleavage in their extracellular domain by proprotein convertases of the subtilisin/kexin family. This cleavage is conserved, not only in different α chains but also across species, suggesting that it might be of functional importance. However, the role of the cleavage in integrin function remains unclear. Initial studies, using site-directed mutagenesis of the cleavage site, demonstrated that uncleaved αIIbβ3 and α4β1 integrins are able to mediate cell adhesion to their respective ligands. Nevertheless, there is now mounting evidence that the post-translational processing of the α chain is essential for the signaling function of, at least, α6β1 and αvβ5 integrins. The absence of cleavage has important consequences on signal transduction pathways leading to alterations in integrin function, such as adhesion or migration, and contributes to the malignant phenotype of tumour cells *in vivo*

Keywords: Proprotein convertases, integrins, cancer

1. CONVERTASES INTRODUCTION/STRUCTURE OF INTEGRINS

Integrins are heterodimeric adhesion molecules that link the extracellular matrix to the actin cytoskeleton and signaling cascades. Integrins receive signals from other receptors, leading to the activation of ligand binding (inside-out signaling) and to matrix assembly. Upon binding ligands, integrins also activate intracellular signaling pathways (outside-in signaling). In this way, cell adhesion is coordinated with other events to orchestrate complex cell behaviors, such as cell migration and/or invasion.

A-Majid Khatib (ed.), Regulation of Carcinogenesis, Angiogenesis and Metastasis by the Proprotein Convertases, 107–119.
© 2006 *Springer.*

CHAPTER 6

In mammals, the integrin receptor family includes at least 18 different α subunits and 8 β subunits that can associate to form 24 distinct integrins. As both α and β subunits recognize the ligand, each αβ association presents its own ligand specificity. In a structural feature, both subunits present a large extracellular domain, a transmembrane segment and a short intracellular region; except for β4 the only subunit bearing a large cytoplasmic domain.

Since 1987 when Hynes spoke about integrin for the very first time [1], a lot of structural information has been collected (for recent reviews, see [2–4]. Electron microscopy for α5β1 and αIIbβ3 [5–8], nuclear magnetic resonance [9] or the mapping of the epitopes structures of conformation sensitive and activating antibodies lead to a global structural model. However, the major advance in our understanding of the relation between structure and function of integrins was provided by the X-ray crystal structure of the extracellular regions of αvβ3 integrin [10, 11].

Integrin α subunits are separated into 2 categories: α subunits that undergo an endoproteolytic processing, like αvβ3, and α subunits bearing an inserted (I) domain, like α1β1 (Figure 1). The αvβ3 integrin presents 12 domains that form an ovoid head and two tails or legs. In the crystal, αvβ3 is severely bent, reflecting an important flexibility that may be linked to integrin regulation [10]. In the bent state,

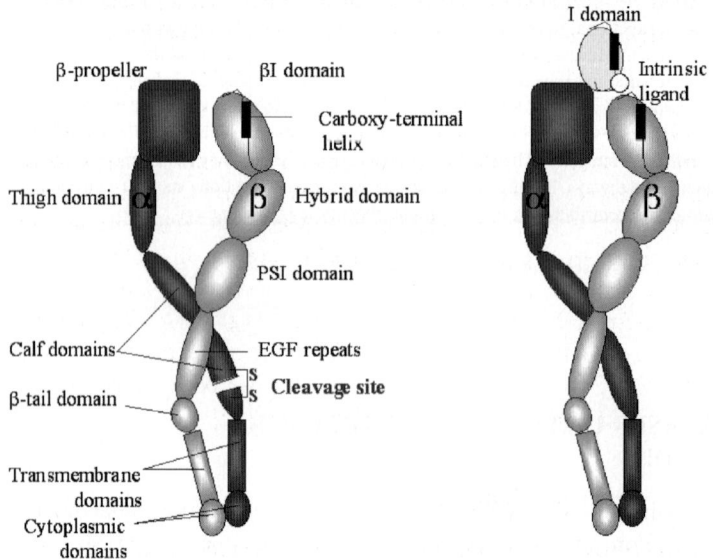

Cleavable integrins   Non cleavable integrins
(α3, α5, α6, α7, α8, α9, αv, αIIb)   (α1, α2, α10, α11, αL, αD, αM, αX)

Figure 1. Comparative structure of integrins with cleaved (left) or uncleaved (right) subunit. Modified from réfs [2, 3]

the ligand-binding domains in the headpiece would be closely juxtaposed to the membrane-proximal portions of the stalks. There is now increasing evidences that conformational changes are propagated all along the molecule and that interactions between the α and β cytoplasmic tails are critical for stabilizing the association between α and β extracellular regions. Disruption of these intracellular interactions by inside-out signals leads to extension of the integrin, repositioning the head region to point away from the cell surface [7].

Half of the 18 integrin α subunits contain an additional I domain inserted into the β-propeller. Where present, this domain is the major site of ligand binding. It is now known how the I domain interacts with ligands. The interaction is due to cooperation between the I domain on α subunit (αI) and an I-like domain located on β subunit (βI domain). The βI domain contains a metal ion-dependent adhesion site (MIDAS) positioned to participate in a ligand interface with αI domain. The αI domain can adopt two stable conformations (closed and open) leading, respectively, to low or high affinity state of the integrin [12, 13].

1.1 Consensus Site for Endoproteolytic Processing of Integrin α Subunit

Among the different integrin α subunits identified to date, α3, α4, α5, α6, α7, α8, α9 αE, αv and αIIb subunits are proteolytically cleaved during their biosynthesis. For the majority of α subunits, the maturation occurs near the carboxyl-terminal part of the extracellular domain, resulting in a heavy chain (about 125 kDa) that is disulphide-linked to a membrane spanning light chain (about 25 kDa). In contrast, the α4 subunit can be expressed at the cell surface either intact or cleaved near the middle of the molecule into non disulphide-linked fragments of 80 and 70 kDa. In the case of αE, the cleavage take place near the NH_2-terminal instead of the COOH-terminal as for other α subunits [14]. The sequence of cleavage regions suggests the existence of two groups that present distinct consensus cleavage sites: Arg-X-(Lys/Arg)-Arg ↓ for α3, α6, α7, α9, αE (group I) and His-X-X-X-(Lys/Arg)-Arg ↓ for α4, α5, α8, αv, and αIIb (group II) (Figure 2).

The presence of an arginine at P4 position or a histidine at P6 position is thus a distinguishing feature of each group. The αIIb subunit has been classed in the group II, although it exhibits both Arg at P4 and His at P6 [15].

1.2 Cleavage Across the Evolution

Integrins are present across the evolution from the nematode *Caenorhabditis. elegans* to mammals. At least three *Drosophila melanogaster* integrins α subunits (αPS1, αPS2 and αPS3) present an endoproteolytic cleavage leading to the formation of a light and a heavy chain linked by a disulphide bond. In addition, *C. elegans* α1 and α2 subunits are also cleaved [16].

Phylogenetic studies of integrin α subunits have shown that I domain-containing subunits form a monophyletic group apart from cleavable α subunits [16, 17].

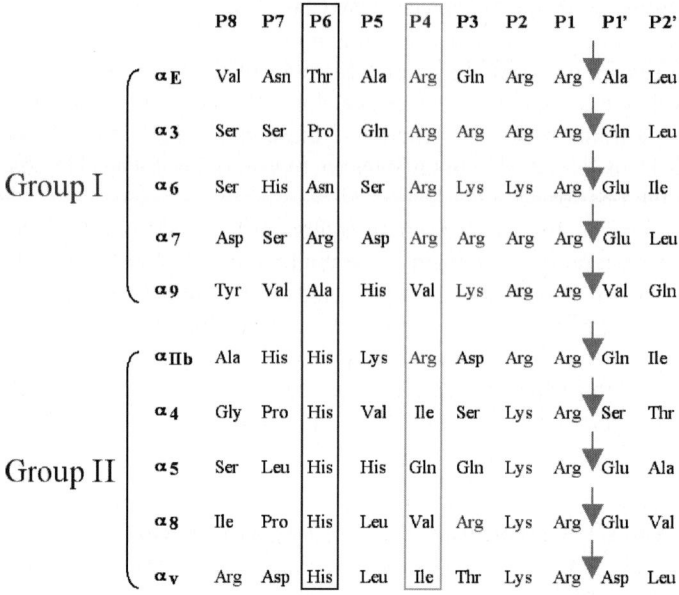

		P8	P7	P6	P5	P4	P3	P2	P1	P1'	P2'
Group I	αE	Val	Asn	Thr	Ala	Arg	Gln	Arg	Arg	Ala	Leu
	α3	Ser	Ser	Pro	Gln	Arg	Arg	Arg	Arg	Gln	Leu
	α6	Ser	His	Asn	Ser	Arg	Lys	Lys	Arg	Glu	Ile
	α7	Asp	Ser	Arg	Asp	Arg	Arg	Arg	Arg	Glu	Leu
	α9	Tyr	Val	Ala	His	Val	Lys	Arg	Arg	Val	Gln
Group II	αIIb	Ala	His	His	Lys	Arg	Asp	Arg	Arg	Gln	Ile
	α4	Gly	Pro	His	Val	Ile	Ser	Lys	Arg	Ser	Thr
	α5	Ser	Leu	His	His	Gln	Gln	Lys	Arg	Glu	Ala
	α8	Ile	Pro	His	Leu	Val	Arg	Lys	Arg	Glu	Val
	αv	Arg	Asp	His	Leu	Ile	Thr	Lys	Arg	Asp	Leu

Figure 2. Alignment of consensus cleavage sites in integrin α subunits from mammals

The I domain family forms two main clusters that contains no known invertebrate members. Cleavable α subunits are gathered into four major clusters. The PS1 cluster contains *Drosophila* αPS1, a *C. elegans* sequence and vertebrate α3, α6 and α7. The PS2 cluster includes *Drosophila* αPS2, a *C. elegans* sequence, two echinoderm sequences and vertebrate α5, α8, αv and αIIb. Interestingly, vertebrate α subunits of the PS1 and PS2 clusters exhibit, respectively, Arg at P4 position or His at P6 in the cleavage region. The PS3 cluster only contains *Drosophila* sequences, αPS1 and two additional α subunits (αPS4 and αPS5) revealed by complete sequencing of the genome. The α4 and α9 subunits form the fourth cluster of α integrins lacking I domain. The phylogeny thus strongly suggests that the common ancestor of deuterostomes (including vertebrates) and protostomes (including insects) possessed at least 3 types of α subunits corresponding to αPS1, αPS2 and αPS3, the latter being lost in the vertebrate lineage [17]. It is worth noting that invertebrates possess only cleavable α subunits and that only vertebrates display α subunits having the I domain. The cleavage of α subunits was thus maintained during evolution from *C. elegans* to vertebrates, while the I domain appeared before the divergence of birds and mammals. Because, except for αE, the presence of cleavage and I domain on integrin α subunits are mutually exclusive, the I domain possibly appeared in place of cleavage. Knowing the importance of the I domain in integrin function, these observations strongly suggest that the ancestral cleavage was maintained for functional interest.

1.3   Integrin Cleavage: Which Enzyme? Which Location?

In our laboratory, we studied the location of α6 subunit cleavage. We showed that endoproteolytic cleavage only occurred after integrin heterodimerisation, probably in the trans-Golgi network (TGN) [18]. Some data are in favor of this hypothesis. First, blocking pro-α6 in the ER by brefeldin A treatment completely prevented its endoproteolytic maturation, but not its association with β4. Second, we have demonstrated that α6 is cleaved after the acquisition of endoglycosidase H resistance (taking place in the cis-Golgi compartment), but before the subunit reached the cell surface [18].

The proprotein convertases (PCs) are a family of endoproteases that activate proproteins by cleavage at basic motifs. Seven PCs are now identified: furin, PC1 (also named PC3), PC2, PACE4, PC4, PC5 (PC6), PC7 (PC8, LPC). The list of substrates activated by these convertases is very vast and includes neuropeptides, peptide hormones, growth and differentiation factors, receptors, adhesion molecules, etc. Observations realised in LoVo cells pointed out the proprotein convertase furin as a possible candidate for integrin cleavage. Indeed, in these cells the gene encoding for furin has a frameshift mutation within one allele and a missense mutation (W547R) within the second allele, resulting in lack of processing activity of the mutant furin [19]. In these furin-deficient cells, integrin subunits α3, α6 and αv are not cleaved and this defective pro-α chain processing was rescued by recombinant furin [20].

To answer the question of the cleavage specificity of α subunits by different PCs, Lissitzky and collaborators analyzed the processing of pro-α integrin subunits by restoring furin activity and over-expressing the other convertases in furin-deficient LoVo cells. They conclude that only furin, PC5A and, to a lesser extent, PACE4 are able to carry out endoproteolytic processing of α3, α5, α6 and αv subunits [21]. An *in vivo* study from Stawowy and collaborators confirmed that PC5 is involved in αv processing. Indeed, on vascular smooth muscle cells, αv and PC5 are co-localised and regulated in the same manner during vascular remodeling [22]. In addition, furin was also found associated with αv subunit in renal podocytes [23]. Results obtained for α4β1 integrin are quite the same that for Lissitzky study. Using LoVo cells, authors showed that α4 is processed by PC5A, furin or PACE4, but not by PC7. However, this leukocyte subunit is quite particular, as its cleavage site is positioned at the middle of the extracellular part of the subunit. Moreover, the integrin can be addressed at the cell surface cleaved or uncleaved and the cleavage status can be correlated with lymphocyte activation. It is one of the few example for direct correlation between α subunit cleavage and integrin function [15].

Others proteases could also cleaved αv integrin subunit if PCs are deficient. Strongin and collaborators used a synthetic convertase inhibitor in breast carcinoma MCF7 cells co-expressing αvβ3 integrin and MT1-MMP. They demonstrated that in these conditions, MT1-MMP is capable of processing the precursor of integrin αv subunit [24]. This cleavage appears to occur between Cys^{852} and Cys^{904} and generates a functional αvβ3 integrin in term of cell adhesion. MT1-MMP is also able to cleave other PC-cleavable α subunits, such as α3 and α5, but failed in the processing of PC-resistant α2 integrin chain [25].

1.4   Integrin Function in Cancer

Cell adhesion and migration are essential for tumour invasion. The interest of slowing down tumour growth by integrin-dependent stabilization of cell-substrate interactions on fibronectin was recognised in the 1980s. Since these initial reports, integrin up- or down-regulation has been regularly observed in tumour progression (for reviews see [26–30]). Integrins act at different levels during tumour progression:
1) Signalling for the lost of neighbouring cell contacts. In addition to classical cell-cell receptors, such as E-cadherin, several integrins are also involved in this process. Thus, β1 overexpression leads to disruption of adherens junctions in epithelial cells [31]. Reciprocally, β1-blocking antibodies induce reformation of adherens junctions and polarisation of tumour cells [32]. Moreover, functional interactions between E-cadherin and αv-containing integrins has been reported in carcinoma cells [33].
2) Migration across the interstitial stroma. Integrins play a key role in the complex process of cell migration. In addition to allowing anchorage necessary for cell displacement, integrins induce signalling ending up to cytoskeleton remodelling. These signalling pathways typically lead to Rho GTPases activation *via* the phosphorylation of focal adhesion kinase (FAK) [34, 35].
3) Survival in the interior circulation (blood or lymph) by adhesion to platelets [36] or lymphocytes [37].
4) Angiogenesis of distal tumour. αvβ3 and αvβ5 integrins have been involved in growth factor- and tumour-induced angiogenesis in multiple animal models (for review [38]). In addition to these largely studied integrins, other members of the family clearly contribute to blood vessel formation, either during development or tumour progression. Thus, mice carrying a targeted deletion of the signalling portion of the integrin β4 subunit display drastically reduced angiogenesis in response to bFGF in Matrigel plug assay [39]. The α1β1 and α2β2 integrins also regulate angiogenesis induced by VEGF [40, 41].

1.5   Consequences of the Perturbation of Integrin Cleavage on Cancer

The role of endoproteolytic cleavage of integrin α subunits is still unclear, but it may play a role in integrin function. As discussed above, the cleavage is conserved, not only in different α chains but also across species (from invertebrates to mammals), suggesting that it might be of functional importance. Furthermore, post-translational proteolysis is a common mechanism required for the synthesis of many biologically active proteins in bacteria, fungi, yeast, invertebrates and mammals [42].

The cleavage status of integrin α subunits has never been studied *in vivo* except for α6 in differentiating lens fibre cells. Authors reported that expression of the uncleaved form of α6 integrin progressively increased relative to the cleaved form during lens cell differentiation, suggesting that the uncleaved form of α6 integrin may have a unique role in the embryonic lens [43].

We have briefly described above the involvement of integrins in tumour invasion. However, although some cleavable integrins, such as αvβ3, αvβ5 or αvβ6 have a central role in cell migration and angiogenesis, it is impossible to ascertain that cleavable integrins are more important than non-cleavable integrins in these processes and that α subunit cleavage is determinant for tumour progression. To address directly the functional importance of integrin cleavage, different studies have been performed *in vitro*, either by site-directed mutagenesis of integrin α subunits or by convertase inhibition in tumour cells.

First directed mutagenesis studies were performed on αIIb and α4 subunits. Mutating the basic residues of convertase recognition site prevented α subunit cleavage, but uncleaved αIIbβ3 and α4β1 integrins were still able to mediate cell adhesion to their respective ligands [44, 45]. Despite of these disappointing results, some studies have been conducted later with other integrins. To examine the importance of cleavage, Delwel and collaborators introduced mutations in the cDNA encoding the RKKR sequence of the α6 subunit. In the human leukemia cells K562 used in this study, the α6Aβ1 integrin is expressed in a resting inactive conformation and need activation to bind ligand. The absence of cleavage of the α6Aβ1 integrin did not affected cell adhesion to laminin-1 after activation by the anti-β1 stimulatory antibody TS2/16. However, interestingly, when activation was performed with the phorbol ester PMA, only wild-type α6Aβ1 integrin was able to recognise laminin-1. The authors conclude that uncleaved α6Aβ1 is able of ligand binding and can transduce outside-in signals but fails in inside-out signalling as it can not be activated by phorbol ester [46, 47]. These results thus suggest that cleavage may provide for the flexibility required to allow proper conformational changes enabling the affinity modulation of the α6Aβ1 integrin.

The inhibition of PCs by specific inhibitors has also been used as an alternative strategy to elucidate the cleavage role in integrin function. In our laboratory we transfected the human colon adenocarcinoma cell line HT29-D4 with a vector encoding for the convertase inhibitor α1-PDX [48]. Clones of stable transfectants were further selected on the basis of their resistance to *Pseudomonas* Exotoxin A, a toxin activated upon cleavage by convertases. The expression of high levels of α1-PDX inhibitor totally blocked the endoproteolytic processing of all the cleavable integrins subunits (α3, α6 and αv) expressed in these cells. This leads to alterations in integrin function such as cell adhesion [48], cell migration [49], proliferation [50] and metastasis formation [51].

The absence of integrin processing has important consequences on signal transduction pathways initiated by ligation of αvβ5 integrin, leading to a reduced attachment to vitronectin. The reduced cell adhesion is most likely not due to changes in integrin affinity, but certainly reflects the inability of the uncleaved integrin to cluster or to interact with its partners [48]. The difference with Delwel results concerning this point [46, 47] could be due, either to a behaviour proper to each integrin, or to the already activated state of integrins in HT29-D4 cells.

Cells expressing uncleaved αv integrins (PDX39P cells) also display a very motile behaviour on vitronectin when compared to control cells (PDX0 cells) and become able to invade collagen gels [49, 51]. The reduced adhesion due to the absence of proteolytic processing of αv subunit could thus, at least partly, explain the increased motility of PDX39P cells. This is likely facilitated by alterations in cytoskeleton remodelling involving the αvβ5 integrin, the sole receptor for vitronectin in HT29-D4 [49].

Subcutaneous inoculation of cells in nude mice revealed that animals injected with α1-PDX-expressing cells exhibited delayed and lower incidence of tumour development, as well as reduced tumour size [50]. Expression of the convertase inhibitor also slightly decreased the tumorigenicity in immunosuppressed newborn rats (Figure 3) [51]. The reduced size of subcutaneous tumours in response to α1-PDX expression is likely due to differences in cell proliferation [50, 52].

However, in spite of their lower growth rate and the smaller tumours they produced, PDX39P cells exhibit a very aggressive behaviour when injected to immunosuppressed rats [51]. Indeed, the tumours they produce showed morphological evidence of higher local invasiveness and infiltrative pattern and produced about 10 times more metastases than control tumours (Figure 3).

The motile and invasive behaviour likely involves integrin αvβ5 because a function-blocking mAb against the αv subunit was efficient to prevent both *in vivo* invasion and *in vitro* cell motility [51]. Taken together, all these results show that the cleavage of αv subunit is essential for the function of αvβ5 integrin and has a marked impact on integrin-dependent events, especially those leading to cell migration. The molecular mechanism by which the endoproteolytic cleavage of αv subunit affects cell migration and aggressiveness is not precisely understood. Potential hypotheses include alterations in the cellular proteins associated with or regulated by αv integrins. Thus, the unconventional processing of αv by membrane type 1 matrix metalloproteinase (MT1-MMP), and the subsequent generation of the modified αvβ3 integrin, results in enhanced functional activity of the integrin [53, 54]. Moreover, further studies indicate that cells displaying the unconventional αvβ3 integrin and cells treated with the furin inhibitor dec-RVKR-cmk show an increased attachment to type I collagen [55]. In addition, α1-PDX-expressing cells acquire the ability to invade collagen, a ligand for α2β1 integrin [51]. These results are consistent with furin-processed αv controlling the cross-talk between αvβ3 and α2β1 integrins. Along with our own results, this clearly demonstrate that the cleavage of αv subunit and the way it is cleaved can drastically influence the integrin function and consequently the behaviour of malignant cells.

The same strategy was used by others authors to demonstrate more broadly the role of PCs in tumour progression (for more details, see [56–58]). Using HT29 cells, Khatib and collaborators clearly demonstrated the role of convertases on activation of several elements involved in growth and malignant phenotype of tumours, such as the IGF-I receptor, plasminogen activators uPA and tPA and the receptor UPAR, or MT1-MMP [50, 56]. They observed that α1-PDX inhibits the IGF-I receptor processing, resulting in a default in IRS-1 signalling, an inhibition

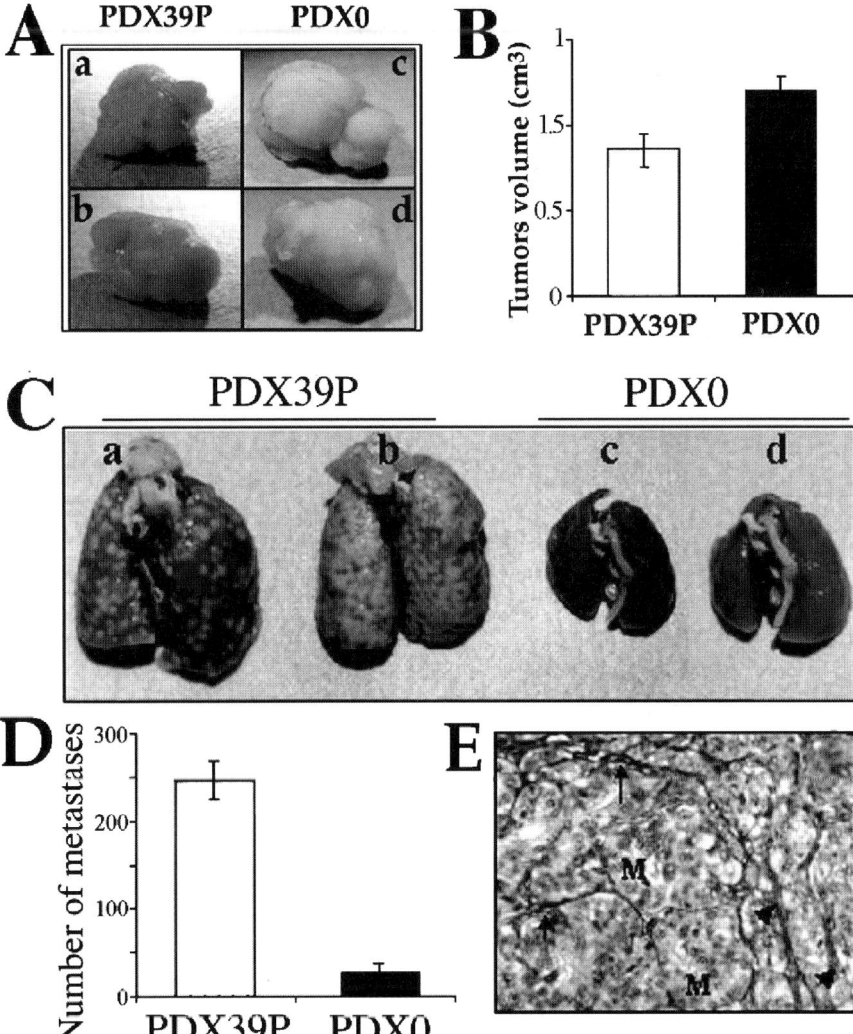

Figure 3. PCs and tumor progression. (**A**) Macroscopic appearance of PDX39P- (**a, b**) and PDX0-induced tumours (**c, d**) three weeks after subcutaneous inoculation of cells in immunosuppressed newborn rats (**B**) Tumours volume was evaluated according to Stragand method Data are mean values ±SD from two experiments (total of 12 rats per cell line). *suite.* 1-PDX expression increases metastases formation. (**C**) Macroscopic appearance of the lungs from PDX39P- (**a, b**) and PDX0-inoculated rats (**c, d**) three weeks after injection of tumour cells. (**D**) Quantification of the number of lung metastases in PDX39P- and control cells-inoculated animals was done on each lung by three experimenters. Data represent mean values ±SD from two experiments (total of 12 rats per cell line). (**E**) Masson trichrome staining of paraffin embedded section. Micrograph (magnification, ×200) shows voluminous metastases (M) with little healthy tissue. Pneumocytes (arrows) are pushed back beneath the pressure of metastatic cells. Reproduced from ref [51], with the authorisation of the American Journal of Pathology

of IGF-1-dependent cell growth and consequently a reduced tumour size [50]. Tumour growth was also reduced in Scid mice inoculated with α1-PDX-transfected astrocytoma cells certainly due to a decrease in tumour cell proliferation [59]. Expression of the inhibitor in neck squamous cell carcinoma cells also reduced *in vitro* cell invasion and *in vivo* invasion across tracheal wall. After injection of these cells in Scid mice, tumours appeared late and remained smaller than with control cells [60].

However, all these results did not take the integrin cleavage status into consideration. Generally, the use of PCs inhibitors is not appropriate to elucidate the role of integrin cleavage on tumour invasion because of the large targets of convertases also involved in this process: metalloproteinases as MT-MMPs, growth factor as TGFβ, growth factor receptors as IGF1-receptor... The αv integrin plays an essential role in tumourigenicity of M21 melanoma cells [61]. We thus transfected both wild-type and mutant αv cDNAs into αv-negative M21-L melanoma cells provided by Dr. David A. Cheresh (Scripps Clinic, La Jolla, USA). Using collagen gel invasion assays, we confirmed that the blockage of αv subunit processing led to an invasive phenotype, as described above for α1-PDX-expressing cells (unpublished results). Overall, our findings indicate that the endoproteolytic processing of αv subunit obviously affects the function of αvβ5 integrin and actively contribute to the malignant phenotype of human tumour cells. The molecular mechanism of the increased motility supported by αv-containing integrins and the aggressiveness that results from it remains to be determined.

1.6   Conclusions

The major conclusions we can draw from the studies described above are that the lack of integrin α subunit cleavage poorly affects the integrin affinity, but strongly alters intracellular signalling. Integrin processing is thus not required for ligand recognition, but may provide for the conformational flexibility required for correct integrin function. Cells bearing uncleaved integrins suffer from adhesion instability problems associated with increased cell migration.

Because of their involvement in the activation of critical proteins implicated in neoplasia, PCs represent interesting potential targets for the development of new therapeutic agents. However, the efficient development and clinical application of PCs inhibitors require a better understanding of their molecular mechanisms *in vitro* and their impact on tumour growth and metastasis in various animal models before testing their efficacy in human disease. Results obtained with uncleaved integrins show that the role of PCs in tumour development and progression is far from being as simple as suggested by previous observations.

REFERENCES

[1] Hynes RO (1987) Integrins: A family of cell surface receptors. Cell **48**:549–554
[2] Carman CV, Springer TA (2003) Integrin avidity regulation: Are changes in affinity and conformation underemphasized? Curr Opin Cell Biol **15**:547–556

[3] Humphries MJ, McEwan PA, Barton SJ, Buckley PA, Bella J, Mould AP (2003) Integrin structure: Heady advances in ligand binding, but activation still makes the knees wobble. Trends Biochem Sci **28**:313–320

[4] Mould AP, Humphries MJ (2004) Regulation of integrin function through conformational complexity: Not simply a knee-jerk reaction? Curr Opin Cell Biol **16**:544–551

[5] Adair BD, Yeager M (2002) Three-dimensional model of the human platelet integrin alpha IIbbeta 3 based on electron cryomicroscopy and x-ray crystallography. Proc Natl Acad Sci U S A **99**:14059–14064

[6] Takagi J, Erickson, HP, Springer TA (2001) C-terminal opening mimics 'inside-out' activation of integrin alpha5beta1. Nat Struct Biol **8**:412–416

[7] Takagi J, Petre BM, Walz T, Springer TA (2002) Global conformational rearrangements in integrin extracellular domains in outside-in and inside-out signaling. Cell **110**:599–511

[8] Takagi J, Strokovich K, Springer TA, Walz T (2003) Structure of integrin alpha5beta1 in complex with fibronectin. Embo J **22**:4607–4615

[9] Beglova N, Blacklow SC, Takagi J, Springer TA (2002) Cysteine-rich module structure reveals a fulcrum for integrin rearrangement upon activation. Nat Struct Biol **9**:282–287

[10] Xiong JP et al. (2001) Crystal structure of the extracellular segment of integrin alpha Vbeta3. Science **294**:339–345

[11] Xiong JP, Stehle T, Zhang R, Joachimiak A, Frech M, Goodman SL, Arnaout MA (2002) Crystal structure of the extracellular segment of integrin alpha Vbeta3 in complex with an Arg-Gly-Asp ligand. Science **296**:151–155

[12] Emsley J, Knight CG, Farndale RW, Barnes MJ, Liddington RC (2000) Structural basis of collagen recognition by integrin α2β1. Cell **101**:47–56

[13] Shimaoka M. et al. (2003) Structures of the alpha L I domain and its complex with ICAM-1 reveal a shape-shifting pathway for integrin regulation. Cell **112**:99–111

[14] Shaw SK, Cepek KL, Murphy EA, Russell GJ, Brenner MB and Parker CM (1994) Molecular cloning of the human mucosal lymphocyte integrin alpha(E) subunit – unusual structure and restricted RNA distribution. J. Biol. Chem. **269**:6016–6025

[15] Bergeron E, Basak A, Decroly E, Seidah NG (2003) Processing of alpha4 integrin by the proprotein convertases: histidine at position P6 regulates cleavage. Biochem J **373**:475–484

[16] Stark KA, Yee GH, Roote CE, Williams EL, Zusman S, Hynes RO (1997) A novel alpha integrin subunit associates with betaPS and functions in tissue morphogenesis and movement during Drosophila development. Development **124**:4583–4594

[17] Hughes AL (2001) Evolution of the integrin alpha and beta protein families. J Mol Evol **52**:63–72

[18] Rigot V, André F, Lehmann M, Lissitzky JC, Marvaldi J, Luis J (1999) Biogenesis of α6β4 integrin in a human colonic adenocarcinoma cell line. Involvement of calnexin. European Journal of Biochemistry **261**:659–666

[19] Takahashi S, Nakagawa T, Kasai K, Banno T, Duguay SJ, Van de Ven WJ, Murakami K, Nakayama K (1995) A second mutant allele of furin in the processing-incompetent cell line, LoVo. Evidence for involvement of the homo B domain in autocatalytic activation. J Biol Chem **270**:26565–26569

[20] Lehmann M. et al. (1998) Deficient processing and activity of type I insulin-like growth factor receptor in the furin-deficient LoVo-C5 cells. Endocrinology **139**:3763–3771

[21] Lissitzky JC, Luis J, Munzer JS, Benjannet S, Parat F, Chrétien M, Marvaldi J, Seidah NG (2000) Endoproteolytic processing of integrin pro-α subunits involves the redundant function of furin and proprotein convertase (PC) 5A, but not paired basic amino acid converting enzyme (PACE) 4, PC5B or PC7. Biochem. J. **346**:133–138

[22] Stawowy P, Graf K, Goetze S, Roser M, Chretien M, Seidah NG, Fleck E, Marcinkiewicz M (2003) Coordinated regulation and colocalization of alphav integrin and its activating enzyme proprotein convertase PC5 in vivo. Histochem Cell Biol **119**:239–245

[23] Mayer G, Boileau G, Bendayan M (2003) Furin interacts with proMT1-MMP and integrin alphaV at specialized domains of renal cell plasma membrane. J Cell Sci **116**:1763–1773

[24] Deryugina EI, Bourdon MA, Jungwirth K, Smith JW, Strongin AY (2000) Functional activation of integrin alpha V beta 3 in tumor cells expressing membrane-type 1 matrix metalloproteinase. Int J Cancer **86**:15–23

[25] Ratnikov BI, et al. (2002) An alternative processing of integrin αv subunit in tumor cells by membrane type-1 matrix metalloproteinase. J Biol Chem **277**:7377–7385

[26] Wehrle-Haller B, Imhof BA (2003) Integrin-dependent pathologies. J Pathol **200**:481–487

[27] Jin H, Varner J (2004) Integrins: Roles in cancer development and as treatment targets. Br J Cancer **90**:561–565

[28] Bogenrieder T, Herlyn M (2003) Axis of evil: molecular mechanisms of cancer metastasis. Oncogene **22**:6524–6536

[29] Hood J, Cheresh D (2002) Role of integrins in cell invasion and migration. Nature Reviews Cancer **2**:91–100

[30] Guo W, Giancotti FG (2004) Integrin signalling during tumour progression. Nat Rev Mol Cell Biol **5**:816–826

[31] Gimond C, van Der Flier A, van Delft S, Brakebusch C, Kuikman I, Collard JG, Fassler R, Sonnenberg A (1999) Induction of cell scattering by expression of beta1 integrins in beta1-deficient epithelial cells requires activation of members of the rho family of GTPases and downregulation of cadherin and catenin function. J Cell Biol **147**:1325–1340

[32] Weaver VM, Petersen OW, Wang F, Larabell CA, Briand P, Damsky C, Bissell MJ (1997) Reversion of the malignant phenotype of human breast cells in three-dimensional culture and in vivo by integrin blocking antibodies. J Cell Biol **137**:231–245

[33] von Schlippe M, Marshall JF, Perry P, Stone M, Zhu AJ, Hart IR (2000) Functional interaction between E-cadherin and alphav-containing integrins in carcinoma cells. J Cell Sci **113**:425–437

[34] Raftopoulou M, Hall A (2004) Cell migration: Rho GTPases lead the way. Dev Biol **265**:23–32

[35] Keely P, Parise L, Juliano R (1998) Integrins and GTPases in tumour cell growth, motility and invasion. Trends Cell Biol **8**:101–106

[36] Felding-Habermann B, et al. (2001) Integrin activation controls metastasis in human breast cancer. Proc Natl Acad Sci U S A **98**:1853–1858

[37] Tantivejkul K, Kalikin, LM, Pienta KJ (2004) Dynamic process of prostate cancer metastasis to bone. J Cell Biochem **91**:706–717

[38] Eliceiri BP, Cheresh DA (2001) Adhesion events in angiogenesis. Curr Opin Cell Biol **13**:563–568

[39] Nikolopoulos SN, Blaikie P, Yoshioka T, Guo W, Giancotti FG (2004) Integrin beta4 signaling promotes tumor angiogenesis. Cancer Cell **6**:471–483

[40] Senger DR, Claffey KP, Benes JE, Perruzzi CA, Sergiou AP, Detmar M (1997) Angiogenesis promoted by vascular endothelial growth factor: Regulation through alpha1beta1 and alpha2beta1 integrins. Proc Natl Acad Sci U S A **94**:13612–13617

[41] Senger DR, Perruzzi CA, Streit M, Koteliansky VE, de Fougerolles AR, Detmar M (2002) The alpha(1)beta(1) and alpha(2)beta(1) integrins provide critical support for vascular endothelial growth factor signaling, endothelial cell migration, and tumor angiogenesis. Am J Pathol **160**:195–204

[42] Denault J-B, Leduc R (1996) Furin/PACE/SPC1: A convertase involved in exocytic and endocytic processing of precursor proteins. FEBS Lett. **379**:113–116

[43] Walker JL, Zhang L, Menko AS (2002) A signaling sole for the uncleaved form of α6 integrin in differentiating lens fiber cells. Dev Biol **251**:195–205

[44] Teixidó J, Parker CM, Kassner PD, Hemler ME (1992) Functional and structural analysis of VLA-4 integrin α4 subunit cleavage. J. Biol. Chem. **267**:1786–1791

[45] Kolodziej M, Vilaire G, Gonder D, Poncz M, Bennett J (1991) Study of the endoproteolytic cleavage of platelet glycoprotein IIb using oligonucleotide-mediated mutagenesis. J Biol Chem. **266**:23499–23504

[46] Delwel GO, Hogervorst F, Sonnenberg A (1996) Cleavage of the α6A subunit is essential for activation of the α6Aβ1 integrin by phorbol 12-myristate 13-acetate. Journal of Biological Chemistry **271**:7293–7296

[47] Delwel G, Kuikman I, van der Schors R, de Melker A, Sonnenberg A (1997) Identification of the cleavage sites in the α6A integrin subunit: Structural requirements for cleavage and functional analysis of the uncleaved α6Aβ1 integrin. Biochemical Journal **324**:263–272

[48] Berthet V, Rigot V, Champion S, Secchi J, Fouchier F, Marvaldi J, Luis J (2000) Role of endoproteolytic processing in the adhesive and signaling functions of αvβ5 integrin. J. Biol. Chem. **275**:33308–33313

[49] Berthet V, Rigot V, Nejjari M, Marvaldi J, Luis J (2004) The endoproteolytic processing of αvβ5 integrin is involved in cytoskeleton remodelling and cell migration. FEBS Lett. **557**:159–163

[50] Khatib AM, Siegfried G, Prat A, Luis J, Chrétien M, Metrakos P, Seidah NG (2001) Inhibition of proprotein convertases is associated with loss of growth and tumorigenicity of HT-29 human colon carcinoma cells: Importance of insulin-like growth factor-1 (IGF-1) receptor processing in IGF-1-mediated functions. J. Biol. Chem. **276**:30686–30693

[51] Nejjari M, et al. (2004) Inhibition of proprotein convertases enhances cell migration and metastases development of human colon carcinoma cells in a rat model. Am. J. Pathol. **164**:1925–1933

[52] Conesa M, Nejjari M, Lissitzky J-C, Luis J (2004) Modulation of integrin function by structural alteration or silencing expression-Potential interest in tumour progression. In: Recent research developments in cell science, Vol. 1, pp. 193–215 Transworld Research Network, Kerala, India

[53] Deryugina EI, Ratnikov B, Monosov E, Postnova TI, DiScipio R, Smith JW, Strongin AY (2001) MT1-MMP initiates activation of pro-MMP-2 and integrin alphavbeta3 promotes maturation of MMP-2 in breast carcinoma cells. Exp Cell Res **263**:209–223

[54] Deryugina EI, Ratnikov BI, Postnova TI, Rozanov DV, Strongin AY (2002) Processing of integrin αv subunit by membrane type 1 matrix metalloproteinase stimulates migration of breast carcinoma cells on vitronectin and enhances tyrosine phosphorylation of focal adhesion kinase. J Biol Chem **277**:9749–9756

[55] Baciu PC, Suleiman EA, Deryugina EI, Strongin AY (2003) Membrane type-1 matrix metalloproteinase (MT1-MMP) processing of pro-αv integrin regulates cross-talk between αvβ3 and α2β1 integrins in breast carcinoma cells. Exp Cell Res **291**:167–175

[56] Khatib AM, Siegfried G, Chrétien M, Metrakos P, Seidah NG (2002) Proprotein convertases in tumor progression and malignancy: Novel targets in cancer therapy. Am J Pathol **160**:1921–1935

[57] Bassi DE, Mahloogi H, Klein-Szanto AJ (2000) The proprotein convertases furin and PACE4 play a significant role in tumor progression. Mol Carcinog **28**:63–69

[58] Taylor NA, Van De Ven WJ, Creemers JW (2003) Curbing activation: Proprotein convertases in homeostasis and pathology. Faseb J **17**:1215–1227

[59] Mercapide J, Lopez De Cicco R, Bassi DE, Castresana JS, Thomas G, Klein-Szanto AJ (2002) Inhibition of furin-mediated processing results in suppression of astrocytoma cell growth and invasiveness. Clin Cancer Res **8**:1740–1746

[60] Bassi DE, Lopez De Cicco R, Mahloogi H, Zucker S, Thomas G, Klein-Szanto AJ (2001) Furin inhibition results in absent or decreased invasiveness and tumorigenicity of human cancer cells. Proc Natl Acad Sci U S A **98**:10326–10331

[61] Felding-Habermann B, Mueller BM, Romerdahl CA, Cheresh DA (1992) Involvement of integrin alpha V gene expression in human melanoma tumorigenicity. J Clin Invest **89**:2018–2022

CHAPTER 7

GROWTH FACTORS: TO CLEAVE
OR NOT TO CLEAVE

ABDEL-MAJID KHATIB[1] AND SIEGFRIED GERALDINE[2]

[1] INSERM, U 716, Equipe AVENIR, Institut de Génétique Moléculaire, Paris, 75010, France;
 Université Paris 7, Paris 75251, France
[2] INSERM U770, 84 Rue Général Leclerc, 94276 Le Kremlin-Bicêtre, France

Abstract: The Majority of growth factors are synthesized as proproteins and are proteolytically processed by the proprotein convertase-like enzymes. The cleavage of some of these molecules is crucial for the mediation of their functions, whereas the unprocessed forms of other growth factors are biologically active and in certain cases oppose the biological action of their processed forms through specific receptors. Here we review the recent advance in our understanding of the importance of growth factors processing by the proprotein convertases on the malignant phenotypes of tumor cells, tumorigenesis and angiogenesis, and the potential use of their maturation blockade/activation as a new potential therapeutic strategy

Keywords: Proprotein convertases, Growth factors

1.   INTRODUCTION

Growth factors are a large family of polypeptide molecules that regulate cell division in many tissues by autocrine and/or paracrine mechanisms. Accumulated reports have demonstrated that these molecules do not only play an important role in embryonic development and adult tissue homeostasis, but are also involved in various pathological situations like infection, wound healing, and tumorigenesis. Many of these molecules are synthesized as precursor proteins and processed and/or activated by the proprotein convertase (PC)-like enzymes [1–3]. Thereby, in certain cases parallel increased expression of both, growth factors and PC may result in tumor growth progression and metastasis through various mechanisms including cell proliferation, survival and invasion. It is therefore possible that inhibition of the processing and the activation of several growth factors by specific and

121

A-Majid Khatib (ed.), Regulation of Carcinogenesis, Angiogenesis and Metastasis by the Proprotein Convertases, 121–135.

potent PC inhibitors might be used against various malignancies induced by these molecules [1, 2].

To date the amino acid (aa) sequence of many growth factors were reported to contain PC-cleavage sites, whereas only few of them were proven to be PC substrates [1–4]. The cleavage of some of these molecules was found to be crucial for the mediation of their functions such as the platelet-derived growth factors (PDGF) [1] and vascular endothelial growth factor-C (VEGF-C) [2], whereas the precursor forms of other growth factors like fibroblast growth factor-23 (FGF-23] are active and their processing by the PC inhibits their activity [5] or mediate the opposite biological action induced by their processed form through specific receptors like the neurotrophins [3].

2.   PROPROTEIN CONVERTASES (PC)

The PCs constitute a family of nine secretory serine proteases. Two are non-basic specific convertases and includes SKI-1 [6] and NARC-1/PCSK9 [7]. The convertase SKI-1 recognizes the motif (R/K)-X-(L,V)-Z, where X is any aa and Z is any aa except Pro, Cys, Glu, and Val [8], and NARC-1/PCSK9 recognizes the sequence VFAQ with Val at the critical position P4 [9]. The other seven convertases are dibasic-specific, namely Furin, PC1, PC2, PC4, PACE4, PC5 and PC7 [4, 6]. These enzymes are implicated in the processing of multiple protein precursors, including proteases, growth factors, and receptors at multibasic recognition sites exhibiting the general motif $(K/R)-(X)_n-(K/R)$ where X is any aa except Cys and $n = 0, 2, 4$ or 6 [1–8].

Here we review the role of these convertases on growth factors processing, activation and inhibition and their impact on growth factors induced-malignant phenotypes of tumor cells, tumorigenesis, and/or angiogenesis.

3.   ACTIVATION OF GROWTH FACTORS BY THE PROPROTEIN CONVERTASES

Of the growth factors produced as precursors and proteolytically activated by the PCs are the PDGFs, IGFs, PTH and endothelins.

3.1   Maturation of PDGFs

PDGFs are a pleotrophic family of growth factors that stimulate various cellular functions including growth, proliferation, and differentiation. The expression of PDGFs has been demonstrated in a number of different solid tumors, from glioblastomas to colon carcinomas [10–13]. In these various tumor types, the biologic role of PDGF signaling can vary from autocrine stimulation of cancer cell growth to paracrine interactions involving adjacent stroma and vasculature leading to tumor progression, angiogenesis and metastasis [10–13].

PDGF are a disulfide-linked dimers composed of two polypeptide chains, denoted A and B and represented *in vivo* by the three PDGFs: PDGF-AA, PDGF-AB and

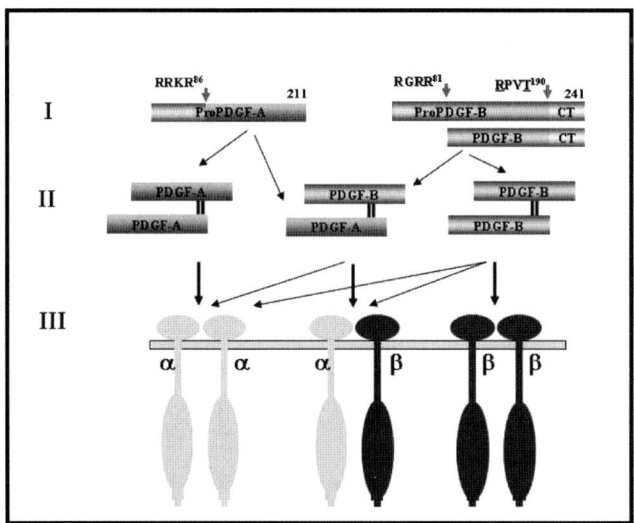

Figure 1. Processing of A- and B-chains of PDGF and activation of their corresponding receptors. The A and B chains are produced as precursor molecules (i) that form disulfide-bonded dimers, undergo proteolytic processing (ii) and induce dimerization of alpha and/or beta receptors (iii)

PDGF-BB [14, 15] that mediate their signals through activation of two tyrosine kinase receptors, PDGFR-α and -β (Figure 1).

Recently two new members of the PDGF family, PDGF-C and PDGF-D were reported. PDGF-CC act as a ligand for PDGFR-α [16] and PDGF-DD as a ligand for PDGFR-β [17]. Of the PDGFs proven to be activated by the PC are PDGF-A [1] and PDGF-B [18]. PDGF-C and PDGF-D were found also to possess potential PC cleavage site [4], however, their processing by the PCs is not yet proven experimentally.

3.1.1   ProPDGF-A processing in the malignant phenotype of tumor cells

Following dimerization of PDGF-A monomers in the ER into a ∼50 kDa form, this complex transits through the Golgi apparatus towards the *trans* Golgi Network (TGN) where it is proteolytically cleaved by the PCs and secreted as ∼30 kDa dimeric product [1]. Although the Furin is the major convertase that process significantly pro-PDGF-A [1], under steady-state conditions, the convertases PC5, PACE4 and PC7 were also found able to mediate pro-PDGF-A processing [1]. Like various PC substrates, PDGF-A processing is inhibited by the naturally occurring PC inhibitors PC-prosegments of Furin (ppFurin), ppPC5, ppPACE4, and by the general PC inhibitors: the Furin-motif variants of α2-macroglobulin (α2–MGF) and serpinα1-antitrypsin (α1-PDX) [1].

Like the other PDGFs, the binding of PDGF-A ligand to its receptor results in the autophosphorylation of the latter [1]. In turn, the PDGF receptor activates an enzyme cascade that includes various phosphorylating enzymes including PKC,

Ras, Raf, and MAPK [1]. Comparative analysis between wild type and mutant (at its PC site) PDGF-A revealed that the unprocessed PDGF-A lost its ability to mediate tyrosine phosphorylation of PDGF-A receptor and were unable to stimulate [^{3}H]thymidine incorporation *in vitro*. *In vivo* experiments revealed that the injection of tumor cells expressing the unprocessed PDGF-A inhibited tumor growth progression [1]. This *in vivo* tumor growth inhibition by the mutated PDGF-A, could be explained by the potential action of the unprocessed proPDGF-A as a dominant negative [19]. Indeed, like the other PDGF ligands, interaction of PDGF-A with their receptors induces the dimerization of the receptor subunits and the formation of PDGFαR-PDGFαR homodimers [1]. The absence of fully processed PDGF-A may affect the dimerization of the corresponding receptors leading to a loss of biological activity. Also, the possible antagonist role of the PDGF-A mutant that may compete with the active PDGF-A for the PDGF receptors it's not ruled out.

3.1.2   *Proprotein convertases in ProPDGF-B processing and secretion*

Contrary to PDGF-A, ProPDGF-B is processed into a major cell surface-associated product of 24 kD and minor secreted product of 30 kD [18]. The ProPDGF-B newly synthesized as 31 kD monomers is rapidly dimerized in the ER via disulfide bridge bonds to yield a 56–60 kD proPDGF-B. The latter is transferred thereafter to the Golgi complex for proteolytic processing to produce an intermediate 40 kD and a final 27 kD product [18]. These forms of PDGF-B are the results of the processing action of Furin, PACE4, PC5, or PC7 that occurs in the N termini at pairs of basic residues RGRR^{81}, followed by a second cleavage by unknown enzyme at the C termini at the surrounding amino acid sequence ARPVT^{19} [18]. This stepwise processing of PDGF-B is controlled by the processing at the PCs site RGRR^{81} as revealed by expression experiments with proPDGF-B mutated at its PC-cleavage site [RGRR^{81} into AGRA^{81}). This mutation eliminated completely the processing of PDGF-B at both the N terminal and C terminal sites. The use of wild-types and mutants PDGF-A and/or PDGF-B chains cDNAs, revealed that the formation of all mature PDGFs isoforms, PDGF-AA, PDGF-BB and PDGF-AB require the processing of both the A and B chains [18]. This processing is also inhibited by α1-PDX and α2–MG-F and some endogenous PC inhibitors [18]. In contrast, the introduction at the second cleavage site of PDGF-B various mutations surrounding the amino acid sequence ARPVT^{190} that contain amino acid sequences bearing a resemblance to potential PC sites failed to block proPDGF-B processing.

The cleavage of PDGF-B at its second site seems to play an important role in the secretion of mature PDGF-B. Indeed, contrary to PDGF-A, the majority of the newly produced PDGF-B is retained at the cell surface, due to the presence of basic residues localized at the C-terminal end of the precursor within the segment aa 212–226. This PDGF-B cell retention motif is localized just after the second cleavage site of PDGF-B (ARPVT^{190}). Mutation introduced at these areas reduced mature PDGF-B secretion, whereas overexpression of Furin resulted in an enhancement of the level of secreted PDGF-B [18]. Thus the processing of PDGF-B at its second site facilitates its secretion thereby the cleaved PDGF-B becomes more diffusible and may act on cells at some distance from the producer cell (Figure 2).

The attachment of the PDGF-B at the cell surface mediated by the residues 212–226 aa of the C-terminal end is due to its interaction with various components of the extracellular matrix, including heparan sulfate, thought to be its major interactor [20, 21]. This would result in the retention of the unprocessed PDGF-B at the cell surface, to be released in a controllable manner by a second processing event. Conversely, several studies reported that deletion of the retention motif in PDGF-B leads to its increased secretion and accumulation in the cell culture media [20, 21]. In addition, the reduced secretion of PDGF-B was previously reported to

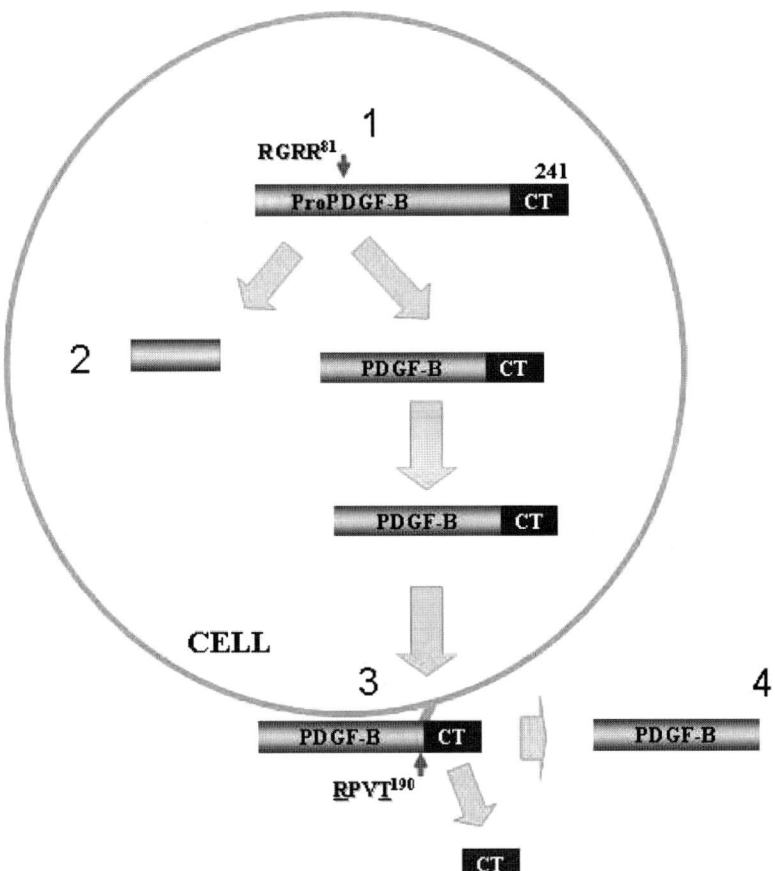

Figure 2. PDGF-B processing and secretion. The newly produced PDGF-b is retained at the cell surface, due to the presence of basic residues localized at the c-terminal (CT) end of the precursor. This PDGF-B cell retention motif is localized just after the second cleavage site of PDGF-B (ARPVT[190]). The processing of PDGF-B at this second site facilitates its secretion thereby the cleaved PDGF-B becomes more diffusible and may act on cells at some distance from the producer cell

limit the action range of PDGF-B *in vivo*, as suggested from experiments with trans-planted keratinocytes expressing wild type or retention motif-truncated PDGF-B into athymic mice [22].

3.2   Maturation of VEGF-C by the Proprotein Convertases in Tumor Angiogenesis

VEGF-C is secreted as a disulfide-bonded homodimer that is processed from the precursor polypeptide at the PC cleavage site HSIIRR^{227}SL (Figure 3). This processing is mediated mainly by the convertases Furin, PC5, and PC7 dividing it into N-terminal (31 kDa) and cysteine-rich C-terminal (29 kDa) polypeptides [2]. Cellular coexpression and *in vitro* kinetics analysis revealed that Furin and PC5 are the most effective proVEGF-C convertases [2]. After Pro-VEGF-C cleavage by the PC an additional processing removes the N-terminal propeptide and generates the 21-kDa VEGF-C (Figure 3).

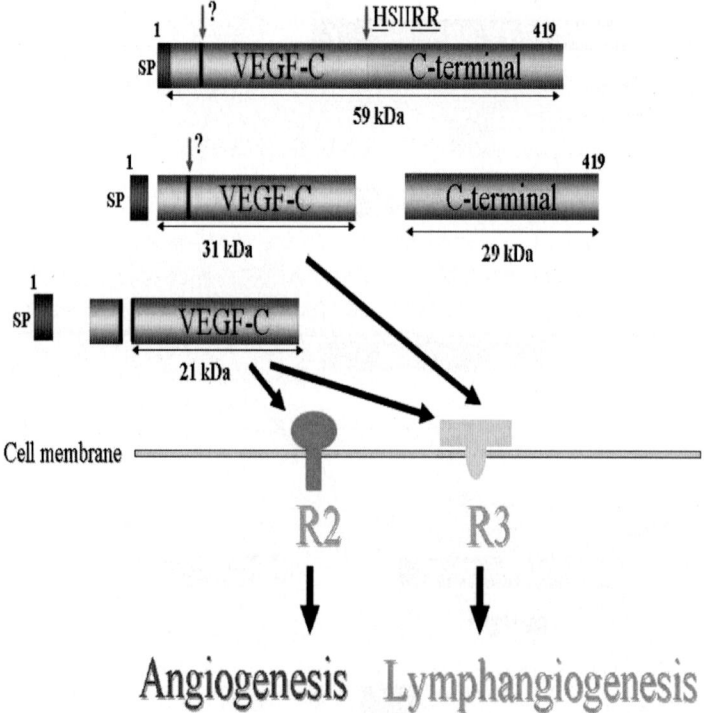

Figure 3. Schematic representation of the primary structure of the 419-AA human proVEGF-C and its receptors R2 and R3. Shown are the signal peptide (SP), PC-processing site (HSIIRR^{227} SL), an unknown protease site (indicated by question mark) that generates the 21-kDA VEGF-C. Activation of VEGFR-2 induces blood vessel formation and activation of VEGFR-3 induces lymphatic vessels formations

The protease (s) that mediates this conversion occurring extracellularly and/or at the cell surface is still unknown. The intracellular proteolytic cleavage of pro-VEGF-C is not a prerequisite for VEGF-C secretion as revealed by the accumulation of proVEGF-C in media derived from the Furin deficient cells LoVo transfected with VEGF-C and following the expression of PC inhibitors in various cells [2]. Of the PC inhibitors that were found to inhibit the processing of VEGF-C the prosegments, ppFurin, ppPC5, ppPACE4, and the serpins α1-PDX, and α2-MG-F [2]. VEGF-C overexpression was linked to various cancers and metastasizing tumors in which VEGF-C was shown to mediate vessel formation [1, 23–25].

Although, it was originally described as a specific growth factor for lymphatic vessels and a ligand for the lymphatic endothelial receptor VEGFR-3 (Flt4), VEGF-C was found also to bind and activates VEGFR-2, which is the major mitogenic signal transducer for VEGF in blood vessel endothelial cells [23–25] (Figure 3). Recently, expression of VEGF-C in CHO cells, failed to stimulate tumor growth *in vitro* and *in vivo* [2], but stimulated both tumor lymphangiogenesis and angiogenesis [2].

The inability of VEGF-C to stimulate cell proliferation is probably due to its weak mitogenic action as compared with the other VEGFs. Indeed, VEGF-C is 50- to 100-fold less potent than VEGF in inducing proliferation of endothelial cells [23–25]. In contrast, VEGF-C is a potent angiogenic factor, and acts synergistically with other growth factors to mediate angiogenesis [25]. Blockade of VEGF-C processing by mutagenesis generates predominately the 59-kDa proVEGF-C form [2]. This form was reported to behave as an antagonist of VEGFR-2 and VEGFR-3 [1, 21–25], and to possibly prevent their optimal oligomerization, which is needed for their signaling functions. Accordingly, the observed reduction in tumor growth, angiogenesis, and lymphangiogenesis of the 59-kDa proVEGF-C–producing tumors [2] suggested to be related to the dominant-negative characteristics of this growth factor [2]. Recently, an opposing role of unprocessed and processed PC substrates was reported for other proteins and demonstrated to be related to distinct signaling pathways. This includes pro–nerve growth factor (proNGF) and other neurotrophins [3], suggesting that similar mechanism may explain the divergent action of the unprocessed and processed pro-VEGF-C on tumorigenesis.

3.3   Maturation of IGFs by the Proprotein Convertases

Although the IGF-1 isolated from human serum is 70 amino acids long, after its production as a precursor protein, the processing of pro-IGF-I to mature IGF-I occurs by cleavage within the processing motif PLKPAKSAR^{71} ↓ RSVRAQR^{77} ↓ HT at two sites : Arg71 to generate IGF-I-(1–70) and at Arg77 to produce IGF-I-(1–76) [26, 27]. The mature IGF-1 form is generated from the two different protein precursors; pro-IGF-IA and pro-IGF-IB.

These molecules are the products of a single gene by alternative splicing that circulate in human serum as a peptide consisting of four domains namely B, C, A and D [26–28]. The sequences of IGF-IA and IGF-IB are identical through B, C, A, and D and through the 15 first amino acid of the pro-region or E domain and both possess the same processing site between E and D domains [26–28]. Previously it was reported that when pro-IGF-1 was expressed in RPE.40 cells, a CHO-K1 derivate cells lacking furin activity, the cleavage of pro-IGF-1 occurs at Arg71 but not at Arg77. Similar results were obtained when pro-IGF-1 was expressed in the furin deficient colon cancer cells LoVo, suggesting that furin is required for cleavage at Arg77 but not at Arg71 [29]. Later, Duguay SJ et al., by expressing wild type and mutant forms of Pro-IGF-1 revealed the preference of Furin to Arg77 and the processing of proIGF-1 at Arg71 by the other members of the proprotein convertases [26, 27].

Similarly, The forms of IGF-2 that are most abundant in normal serum are the mature IGF-2 (1–67) and IGF-2-(1–87) [30]. Expression experiments with pro-IGF-2 mutants containing substitutions at potential cleavage sites revealed that pro-IGF-2 is processed at Arg^{104} in HK 293 cells [30]. Coexpression of pro-IGF-2 with furin, PACE4, PC5A, PC5B, or PC7 resulted in enhanced or complete processing of pro-IGF-2 confirming the cleavage of IGF-2 by the PCs. Contrary to IGF-1 of witch the processing occurs only intracellularly, proIGF-2 is processed during transport through the cell and extracellularly [30].

In addition to its critical role in the growth and development of many tissues and the regulation of the overall growth, particularly prenatal one, IGF-1 is also implicated in various pathophysiological conditions, and is thought to play a particularly prominent role in tumorigenesis. Previously, the potent mitogenic activity of IGF-I in cell culture made it an obvious candidate risk factor in cancer development. To date, a significant amount of data had been accumulated suggesting the important role of this growth factor in various cancers including prostate, breast, colon and lung cancers [31–36]. In the prostate, IGF-I is a mitogen for prostate epithelial cells. Various clinical studies revealed strong positive association between IGF-I levels and prostate cancer risk [31]. In breast cancer, also a positive relation between circulating IGF-I concentration and risk of breast cancer was found among premenopausal but not postmenopausal women, suggesting the use of plasma IGF-I concentrations in the identification of women at high risk of breast cancer [32]. Similarly, in colorectal [33] and lung [34] cancers various studies have reported positive associations between serum IGF-I and these cancer risks. Although investigators have compared the activity of mature and unprocessed IGFs in several *in vitro* assays such as thymidine incorporation, to date the importance of IGF-1 and IGF-2 processing by the PCs in the malignant phenotype of tumor cells and tumor progression is still not yet proven experimentally.

3.4   PTH and PTH-rP Processing by the PCs and Tumorigenesis

Both PTH and PTHrP hormones are initially synthesized as larger inactive precursor proteins that undergo processing to release the active molecules [37–42]. While PTH is involved in the regulation of normal extracellular fluid calcium homeostasis and expressed almost exclusively in the parathyroid gland, pro-PTHrP is widely expressed in both none-endocrine and neuroendocrine cells and considered as the major pathogenic endocrine factor in malignancy-associated hypercalcemia and bone metastasis [43]. Usually, primary cancers metastasize to bone by a multistep process that involves interactions between tumor cells and normal host cells. In breast carcinoma, the final step in bone metastasis is mediated by osteoclasts that are stimulated by local production of the tumor peptide PTH-rP. During this step PTHrp mediate bone destruction prior metastasis. In contrast in prostate carcinomas PTHrp stimulates osteoblasts to make new bone.

PTH mRNA, encodes a signal sequence of 25 amino acids and a basic pro-peptide of 6 amino acids [40–42] that constitute the proPTH. The latter is then transported to the *trans*-Golgi network where it's processed and the mature PTH polypeptide of 84 amino acids is packaged into secretory granules. This processing event occurs in the parathyroid chief cell, in the *trans*-Golgi network rather than in secretory granules, which is consistent with processing of proPTH by furin or a furin-like enzyme (PC7) rather than PC1 or PC2. These data are supported by the co-expression of proPTH with furin but not with PC1 and PC2 in the parathyroid cell.

In human, tree mature peptide isoforms derived from pro-PTHrp were previously reported, namely the 139, 141 and 173 aa. These peptides were proposed to be translated from three different mRNAs generated by alternative splicing [38]. These peptide isoformes differ only at their COOH-terminus. The intracellular conversion of pro-PTHrp to the mature products 139, 141 and 173 aa is mediated by furin at the motif Arg-Leu-Lys-Arg. Additional cleavage occurs at these matures peptides with undetermined enzymes to generate the 1–36 peptide able to interact with the receptor. The importance of PTHrP on the malignant phenotype was previously reported by Liu et al.. The authors demonstrated that alteration of PTHRP processing diminish the hypercalcemic endocrine actions of PTHRP and reduce autocrine/paracrine effects of PTHRP on tumor cell growth *in vitro* and *in vivo* as well [44].

3.5   Processing of Endothelin-1 by the Proprotein Convertases

By acting directly on endothelial cells, endothelin-1 (ET-1), was reported to modulate different stages of neovascularization, including proliferation, migration, invasion, proteases production and morphogenesis [45, 46]. ET-1 was found also able to modulate tumor angiogenesis indirectly through the induction of various angiogenic factors such as vascular endothelial growth factor (VEGF) [46]. In addition, tumor cells were reported to express ET-1 receptor, suggesting the potential participation of ET-1 in the tumorigenic and/or metastatic potential of several tumor cells [46].

The processing of proET-1 is a multi-step event that requires the participation of at least three proteases [47–48]. This three-step process consisting of an initial prote-olytic cleavage of the pro-endothelin-1 precursor to big endothelin-1, C-terminal trimming by a carboxypeptidase and further processing of the big endothelin-1 peptide to endothelin-1 by endothelin-converting enzyme (ECE). The conversion of pro-ET-1 to big-ET-1 is mediated by furin or PC7 [47, 48]. Introduction of a point mutation into PC cleavage site of the pro-ET-1 prevent its processing at the Arg-Ser-Lys-Arg motif. Co-transfection with ECE-1 cDNA revealed that the cleavage at Arg52 is not essential for its processing by ECE-1, but cleavage at Arg92 by furin-like convertase is absolutely necessary for cleavage by ECE-1 at Trp73 to produce mature endothelin-1 [47, 48].

4.   INHIBITION OF FGF-23 ACTIVITY BY THE PCS

In our knowledge, to date, FGF-23 is the only growth factor reported to be produced as a biologically active precursor and its activity is reduced or inhibited following its processing by the proprotein convertases. This growth factor was recently reported to be involved in the pathogenesis of two similar hypophosphatemic diseases, autosomal dominant hypophosphatemic rickets/osteomalacia (ADHR) and tumor-induced osteomalacia (TIO) [5, 49–54]. The kidneys of individuals with these disorders have a reduced capacity to reabsorb phosphate that results in phosphate wasting through phosphate excretion. In both cases a highly levels of circulating FGF-23 was reported. Overproduction of FGF-23 by tumors causes TIO. In contrast, ADHR derives from missense mutations in FGF-23 gene [49–54]. Interestingly, analysis of the ADHR derived FGF-23, revealed that FGF-23 of all ADHR individuals contains a mutation in a region that encodes a proprotein convertases motif [49–54]. Previously, this finding raised the possibility that failure in the processing of FGF-23 during its synthesis alter FGF-23 activity. Recent studies revealed that this FGF-23 mutation is responsible for the elevated levels of unprocessed and biologically active circulating FGF-23. Thereby, by preventing FGF-23 cleavage, this mutation seems to somehow stabilize FGF-23 and increases its levels that in turn inhibit reabsorption of phosphates from the kidney into blood leading to ADHR disorders [48–54]. FGF-23 is a 30 kDa secreted protein that is processed by furin into an 18 kDa and 12 kDa fragments. *In vitro* experiments revealed that FGF-23 precursor is processed between Arg179 and Ser180 and when the N-terminal and C-terminal processed fragments of FGF-23 were injected into mice, only unprocessed FGF-23 reduced serum phosphate levels [5, 53]. These studies highlighted the direct implication of FGF-23 processing in the regulation of FGF-23 activity [5, 53, 54].

5.   UNPROCESSED AND PROCESSED NEUROTROPHINS
MEDIATE OPPOSING BIOLOGICAL ACTION

Neurotrophic factors are a group of proteins with a similar structure that consists of nerve growth factor growth factor (NGF), brain-derived neurotrophic factor (BDNF), neurotrophin 3 (NT-3) and neurotrophin 4/5 (NT-4/5) [55]. Initially, neurotrophins were described as important signaling molecules for the survival and neurite outgrowth of neurons [56]. Somewhat surprisingly, there is a growing body of evidence indicating that these molecules are also produced by various cancer cells and are major stimulators of various cancer cells to induce growth and survival [57].

The actions of these neurotrophins are mediated by two different types of signal-transducing receptors [58], first, by activating specific receptor tyrosine kinases of the Trk family [59, 60] and, second, via the p75NTR neurotrophin receptor, which is a member of the tumor necrosis factor receptor family [59, 60].

All neurotrophins are generated from protein precursors that contain at their N-termini hydrophobic signal peptides followed by pro-regions domains. Their mature forms share approximate similar molecular sizes and their primary sequence identities are close to 50% [61]. The intracellular cleavage of the pro-neurotrophins is mediated by Furin, PACE4 and/or PC5/6-B [61]. To produce the active forms of these growth factors the cleavage of their precursors takes place following pairs of the basic amino acids motif Arg-X-(Lys/Arg)-Arg, where X is Ser, Val and Arg for proNGF/proNT-4/5, proBDNF and proNT-3 respectively [61, 62]. Recently, several groups explored the biological functions of precursor forms of neurotrophins [62]. Analysis of human brain revealed the predominance of the Pro-NGF form. Using pro-NGF mutant with altered furin-cleavage site [3] revealed that the unprocessed pro-NGF is able to bind to p75NTR-expressing cells but not to TrkA-expressing cells. Accordingly, this pro-NGF failed to induce neurite outgrowth by PC12 cells expressing TrkA and pro-NGF was found to exert a proapoptotic effect via p75NTR on smooth muscle cells and on corticospinal neurons after adult central nervous system injury. Most recently, Nykjaer and coworkers [63] showed that the high affinity binding of pro-NGF to p75NTR-expressing cells is not mediated solely by p75NTR but requires the presence of an additional pro-NGF receptor called "sortilin/neurotensin receptor-3" [63]. Indeed, in the absence of sortilin, pro NGF bound to either TrkA or to p75NTR with similar low affinity. In contrast, the affinity of mature NGF to TrkA or p75NTR was almost tenfold higher [63, 64]. Similar to pro-NGF, the precursor form of pro-BDNF was also found to be secreted by various cells such as hippocampal neurons [3]. The use of pro-BDNF with mutated furin cleavage site showed the same receptor-interaction profile as mature BDNF, binding to the extracellular domains of TrkB and p75NTR but not to TrkA or TrkC. Accordingly, pro-BDNF induced tyrosine phosphorylation of TrkB in cells expressing TrkB. Mutagenesis studies revealed that residue R103, present in both pro-BDNF and mature BDNF and in all neurotrophins, plays a key role in the binding of these growth factors to TrkB [64].

6. CONCLUSIONS

Cell growth is controlled by a delicate balance between growth-promoting and growth-inhibiting factors. In normal tissue the production and activity of these factors are controlled and regulated to maintain the normal integrity and function of the organ. Using a variety of mechanisms the malignant cells usually escape this control; thereby leading to unregulated and aberrant cell growth. Its now well established that the driver for tumor growth is the aberrant production of various active growth factors and the overexpressing of these molecules by the tumors is associated with poor clinical outcome.

In order to target tumor progression, a variety of strategies were proposed to inhibit the activity of these growth factors, their receptors or signaling pathways. The rapid progress that has been made in our understanding of the role of growth factors processing by the PC in the mediation, inhibition or regulation of their function suggested the potential use of the PC inhibitors as a new tools to regulate their functions. However, further work is required to understand the key mechanisms that determine if the processed form of a growth factor should be active or inactive molecule or mediate the opposite biological action induced by its corresponding unprocessed form. Based on the studies reported for neurotrophins [3], the opposing role of their processed and unprocessed forms seems to involve the activation of different receptors. Although the mechanism is not defined yet, similar scenario is observed for VEGF-C and probably other growth factors. The inhibition of VEGF-C processing revealed to reduce the processes of tumor angiogenesis, and lymphangiogenesis. It will be necessary to investigate whether growth factors of which the processing by the convertases mediates opposing biological actions is due to the activation of specific and different receptors as reported for the neurotrophins [3]. Also, further investigations are required to identify other growth factors that are active under their unprocessed forms and inactive when processed by the PCs.

ACKNOWLEDGEMENT

This work was supported by AVENIR Award, INSERM to AM K, Paris, France.

REFERENCES

[1] Siegfried G, Khatib AM, Benjannet S, Chretien M, Seidah NG (2003) The proteolytic processing of pro-platelet-derived growth factor-A at RRKR (86) by members of the proprotein convertase family is functionally correlated to platelet-derived growth factor-A-induced functions and tumorigenicity. Cancer Res 63:1458–1463

[2] Siegfried G, Basak A, Cromlish JA, Benjannet S, Marcinkiewicz J, Chretien M, Seidah NG, Khatib AM (2003) The secretory proprotein convertases furin, PC5, and PC7 activate VEGF-C to induce tumorigenesis. J Clin Invest 111:1723–1732

[3] Lee R, Kermani P, Teng KK., Hempstead BL (2001) Regulation of cell survival by secreted proneurotrophins. Science 294:1945–1948

[4] Khatib AM, Siegfried G, Chretien M, Metrakos P, Seidah NG (2002) Proprotein convertases in tumor progression and malignancy: novel targets in cancer therapy. Am J Pathol 160:1921–1935

[5] Fukumoto S (2005) Post-translational modification of Fibroblast Growth Factor 23. Ther Apher Dial 9:319–322

[6] Seidah NG, Chretien M (1999) Proprotein and prohormone convertases: a family of subtilases generating diverse bioactive polypeptides. Brain Res 848:45–62

[7] Seidah NG, Prat A (2002) Precursor convertases in the secretory pathway, cytosol and extracellular milieu. Essays Biochem 38:79–94

[8] Seidah NG, Benjannet S, Wickham L, Marcinkiewicz J, Jasmin SB, Stifani S, Basak A, Prat A, Chretien M (2003) The secretory proprotein convertase neural apoptosis-regulated convertase 1 (NARC-1): liver regeneration and neuronal differentiation. Proc Natl Acad Sci USA 100:928–933

[9] Benjannet S, Rhainds D, Essalmani R, Mayne J, Wickham L, Jin W, Asselin MC, Hamelin J, Varret M, Allard D, Trillard M, Abifadel M, Tebon A, Attie AD, Rader DJ, Boileau C, Brissette L, Chretien M, Prat A, Seidah NG (2004) NARC-1/PCSK9 and its natural mutants: zymogen cleavage

and effects on the low density lipoprotein (LDL) receptor and LDL cholesterol. J Biol Chem 279:48865–48875

[10] Heldin CH, Westermark B (1999) Mechanism of action and *in vivo* role of platelet-derived growth factor. Physiol Rev 79:1283–1316

[11] Sulzbacher I, Traxler M, Mosberger I, Lang S, Chott A (2000) Platelet-derived growth factor-AA and -alpha receptor expression suggests an autocrine and/or paracrine loop in osteosarcoma. Mod Pathol 13:632–637

[12] Lokker NA, Sullivan CM, Hollenbach SJ, Israel MA, Giese NA (2002) Platelet-derived growth factor (PDGF) autocrine signaling regulates survival and mitogenic pathways in glioblastoma cells: evidence that the novel PDGF-C and PDGF-D ligands may play a role in the development of brain tumors. Cancer Res 62:3729–3735

[13] Heldin CH, Rönnstrand L (1997) Growth factor receptors in cell transformation. In: Peters G, Vousden K (eds), Frontiers in Molecular Biology: Oncogenes and Tumor Suppressor Genes, Oxford, Oxford University Press, pp 55–85

[14] Hart CE, Bailey M, Curtis DA, Osborn S, Raines E, Ross R, Forstom JW (1990) Purification of PDGF-AB and PDGF-BB from human platelet extracts and identification of all three PDGF dimers in human platelets. Biochemistry 29:166–172

[15] Robbins KC, Antoniades HN, Devare SG, Hunkapiller MW, Aaronson SA (1983) Structural and immunological similarities between simian sarcoma virus gene product(s) and human platelet-derived growth factor. Nature 305:605–608

[16] Li X, Ponten A, Aase K, Karlsson L, Abramsson A, Uutela M, Backstrom G, Hellstrom M, Bostrom H, Li H, Soriano P, Betsholtz C, Heldin CH, Alitalo K, Ostman A, Eriksson U (2000) PDGF-C is a new protease-activated ligand for the PDGF alpha-receptor. Nat Cell Biol 2:302–309

[17] Bergsten E, Uutela M, Li X, Pietras K, Ostman A, Heldin CH, Alitalo K, Eriksson U (2001) PDGF-D is a specific, protease-activated ligand for the PDGF beta- receptor. Nat Cell Biol 3:512–516

[18] Siegfried G, Basak A, Prichett-Pejic W, Scamuffa N, Ma L, Benjannet S, Veinot JP, Calvo F, Seidah N, Khatib AM (2005) Regulation of the stepwise proteolytic cleavage and secretion of PDGF-B by the proprotein convertases. Oncogene 24:6925–6935

[19] Mercola M, Deininger PL, Shamah SM, Porter J, Wang CY, Stiles CD (1990) Dominant-negative mutants of a platelet-derived growth factor gene. Genes Dev 4:2333–2341

[20] LaRochelle WJ, May-Siroff M, Robbins KC, Aaronson SA (1991) A novel mechanism regulating growth factor association with the cell surface: identification of a PDGF retention domain. Genes Dev 5:1191–1199

[21] Rolny C, Spillmann D, Lindahl U, Claesson-Welsh L (2002) Heparin amplifies platelet-derived growth factor (PDGF)- BB-induced PDGF alpha -receptor but not PDGF beta -receptor tyrosine phosphorylation in heparan sulfate-deficient cells. Effects on signal transduction and biological responses. J Biol Chem 277:19315–19321

[22] Eming SA, Yarmush ML, Krueger GG, Morgan JR (1999) Regulation of the spatial organization of mesenchymal connective tissue: effects of cell-associated versus released isoforms of platelet-derived growth factor. Am J Pathol 154:281–289

[23] Joukov V, Pajusola K, Kaipainen A, Chilov D, Lahtinen I, Kukk E, Saksela O, Kalkkinen N, Alitalo K (1996) A novel vascular endothelial growth factor, VEGF-C, is a ligand for the Flt4 (VEGFR-3) and KDR (VEGFR-2) receptor tyrosine kinases. EMBO J 15:290–298

[24] Joukov V, Sorsa T, Kumar V, Jeltsch M, Claesson-Welsh L, Cao Y, Saksela O, Kalkkinen N, Alitalo K (1997) Proteolytic processing regulates receptor specificity and activity of VEGF-C. EMBO J 16:3898–3911

[25] Pepper MS, Mandriota SJ, Jeltsch M, Kumar V, Alitalo K (1998) Vascular endothelial growth factor (VEGF)-C synergizes with basic fibroblast growth factor and VEGF in the induction of angiogenesis in vitro and alters endothelial cell extracellular proteolytic activity. J Cell Physiol 177:439–452

[26] Duguay SJ, Lai-Zhang J, Steiner DF (1995) Mutational analysis of the insulin-like growth factor I prohormone processing site. J Biol Chem 270:17566–17574

[27] Duguay SJ, Milewski WM, Young BD, Nakayama K, Steiner DF (1997) Processing of wild-type and mutant proinsulin-like growth factor-IA by subtilisin-related proprotein convertases. J Biol Chem 272:6663–6670

[28] Daughaday WH, Rotwein P (1989) Insulin-like growth factors I and II. Peptide, messenger ribonucleic acid and gene structures, serum, and tissue concentrations. Endocr Rev 10:68–91

[29] Lehmann M, Andre F, Bellan C, Remacle-Bonnet M, Garrouste F, Parat F, Lissitsky JC, Marvaldi J, Pommier G (1998) Deficient processing and activity of type I insulin-like growth factor receptor in the furin-deficient LoVo-C5 cells. Endocrinology 139:3763–3771

[30] Duguay SJ, Jin Y, Stein J, Duguay AN, Gardner P, Steiner DF (1998) Post-translational processing of the insulin-like growth factor-2 precursor. Analysis of O-glycosylation and endoproteolysis. J Biol Chem 273:18443–18451

[31] Chan JM, Stampfer MJ, Giovannucci E, Gann PH, Ma J, Wilkinson P, Hennekens CH, Pollak M (1998) Plasma insulin-like growth factor-I and prostate cancer risk: a prospective study. Science 279:563–566

[32] Hankinson SE, Willett WC, Colditz GA, Hunter DJ, Michaud DS, Deroo B, Rosner B, Speizer FE, Pollak M (1998) Circulating concentrations of insulin-like growth factor-I and risk of breast cancer. Lancet 351:1393–1396

[33] Probst-Hensch NM, Yuan JM, Stanczyk FZ, Gao YT, Ross RK, Yu MC (2001) IGF-1, IGF-2 and IGFBP-3 in prediagnostic serum: association with colorectal cancer in a cohort of Chinese men in Shanghai. Br J Cancer 85:1695–1699

[34] Lukanova A, Toniolo P, Akhmedkhanov A (2001) A prospective study of insulin-like growth factor-I, IGF-binding proteins-1, -2 and -3 and lung cancer risk in women. Int J Cancer 92:888–892

[35] Holthuizen PE, Cleutjens CB, Veenstra GJ, van der Lee FM, Koonen-Reemst AM, Sussenbach JS (1993) Differential expression of the human, mouse and rat IGF-II genes. Regul Pept 48:77–89

[36] DeChiara TM, Robertson EJ, Efstratiadis A (1991) Parental imprinting of the mouse insulin-like growth factor II gene. Cell 22:849–859

[37] Halloran BP, Nissenson RA (eds) (1992) Parathyroid Hormone-related Protein: Normal Physiology and Its Role in Cancer, CRC Press, Boca Raton, FL

[38] Kemper B, Habener JF, Mulligan R, Potts JT Jr, Rich A (1974) Pre-proparathyroid hormone: a direct translation product of parathyroid messenger RNA. Proc Natl Acad Sci USA 71:3731–3735

[39] Habener JF, Amherdt M, Ravazzola M, Orci L (1979) Parathyroid hormone biosynthesis. Correlation of conversion of biosynthetic precursors with intracellular protein migration as determined by electron microscope autoradiography. J Cell Biol 80:715–731

[40] Canaff L, Bennett HP, Hou Y, Seidah NG, Hendy GN (1999) Proparathyroid hormone processing by the proprotein convertase-7: comparison with furin and assessment of modulation of parathyroid convertase messenger ribonucleic acid levels by calcium and 1, 25-dihydroxyvitamin D3. Endocrinology 140:3633–3642

[41] Mangin M, Webb AC, Dreyer BE, Posillico JT, Ikeda K, Weir EC, Stewart AF, Bander NH, Milstone L, Barton DE et al. (1988) Identification of a cDNA encoding a parathyroid hormone-like peptide from a human tumor associated with humoral hypercalcemia of malignancy. Proc Natl Acad Sci USA 85:597–601

[42] Yasuda T, Banville D, Hendy GN, Goltzman D (1988) Human renal carcinoma expresses two messages encoding a parathyroid hormone-like peptide: evidence for the alternative splicing of a single-copy gene. Proc Natl Acad Sci USA 85:4605–4609

[43] Guise TA (1997) Parathyroid hormone-related protein and bone metastases. Cancer 80:1572–1580

[44] Liu B, Amizuka N, Goltzman D, Rabbani SA (1995) Inhibition of processing of parathyroid hormone-related peptide by anti-sense furin: effect in vitro and in vivo on rat Leydig (H-500) tumor cells. Int J Cancer 63:276–281

[45] Dawas K, Loizidou M, Shankar A, Ali H, Taylor I (1999) Angiogenesis in cancer: the role of endothelin-1. Ann R Coll Surg Engl 81:306–310

[46] Bagnato A, Spinella F (2003) Emerging role of endothelin-1 in tumor angiogenesis. Trends Endocrinol Metab 14:44–50

[47] Blais V, Fugere M, Denault JB, Klarskov K, Day R, Leduc R (2002) Processing of proendothelin-1 by members of the subtilisin-like pro-protein convertase family. FEBS Lett 524:43–48

[48] Kido T, Sawamura T, Hoshikawa H, D'Orleans-Juste P, Denault JB, Leduc R, Kimura J, Masaki T (1997) Processing of proendothelin-1 at the C-terminus of big endothelin-1 is essential for proteolysis by endothelin-converting enzyme-1 in vivo. Eur J Biochem 244:520–526

[49] Larsson T, Davis SI, Garringer HJ, Mooney SD, Draman MS, Cullen MJ, White KE (2005) Fibroblast growth factor-23 mutants causing familial tumoral calcinosis are differentially processed. Endocrinology 146:3883–3891

[50] Fukumoto S, Yamashita T (2002) Fibroblast growth factor-23 is the phosphaturic factor in tumor-induced osteomalacia and may be phosphatonin. Curr Opin Nephrol Hypertens 11:385–389

[51] Yamazaki Y, Okazaki R, Shibata M, Hasegawa Y, Satoh K, Tajima T, Takeuchi Y, Fujita T, Nakahara K, Yamashita T, Fukumoto S (2002) Increased circulatory level of biologically active full-length FGF-23 in patients with hypophosphatemic rickets/osteomalacia. J Clin Endocrinol Metab 87:4957–4960

[52] Shimada T, Muto T, Urakawa I, Yoneya T, Yamazaki Y, Okawa K, Takeuchi Y, Fujita T, Fukumoto S, Yamashita T (2002) Mutant FGF-23 responsible for autosomal dominant hypophosphatemic rickets is resistant to proteolytic cleavage and causes hypophosphatemia in vivo. Endocrinology 143:3179–3182

[53] White KE, Carn G, Lorenz-Depiereux B, Benet-Pages A, Strom TM, Econs MJ (2001) Autosomal-dominant hypophosphatemic rickets (ADHR) mutations stabilize FGF-23. Kidney Int 60:2079–2086

[54] Ito N, Fukumoto S, Takeuchi Y, Yasuda T, Hasegawa Y, Takemoto F, Tajima T, Dobashi K, Yamazaki Y, Yamashita T, Fujita T (2005) Comparison of two assays for fibroblast growth factor (FGF)-23. J Bone Miner Metab 23:435–440

[55] Bibel M, Barde YA (2000) Neurotrophins: key regulators of cell fate and cell shape in the vertebrate nervous system. Genes Dev 14:2919–2937

[56] Lee FS, Kim AH, Khursigara G, Chao MV (2001) The uniqueness of being a neurotrophin receptor. Curr Opin Neurobiol 11:281–286

[57] Denkins Y, Reiland J, Roy M, Sinnappah-Kang ND, Galjour J, Murry BP, Blust J, Aucoin R, Marchetti D (2004) Brain metastases in melanoma: roles of neurotrophins. Neuro-oncol 6.154–165

[58] Bothwell M (1995) Functional interactions of neurotrophins and neurotrophin receptors. Annu Rev Neurosci 18:223–253

[59] Kaplan DR, Miller FD (2000) Neurotrophin signal transduction in the nervous system. Curr Opin Neurobiol 10:381–391

[60] Roux PP, Barker PA (2002) Neurotrophin signaling through the p75NTR neurotrophin receptor. Prog Neurobiol 67:203–233

[61] Mowla SJ, Farhadi HF, Pareek S, Atwal JK, Morris SJ, Seidah NG, Murphy RA (2001) Biosynthesis and post-translational processing of the precursor to brain-derived neurotrophic factor. J Biol Chem 276:12660 6

[62] Chao MV, Bothwell M (2002) Neurotrophins: to cleave or not to cleave. Neuron 33:9–12

[63] Nykjaer A, Lee R, Teng KK, Jansen P, Madsen P, Nielsen MS, Jacobsen C, Kliemannel M, Schwarz E, Willnow TE, Hempstead BL, Petersen CM (2004) Sortilin is essential for proNGF-induced neuronal cell death. Nature 427:843–848

[64] Fayard B, Loeffler S, Weis J, Vogelin E, Kruttgen A (2005) The secreted brain-derived neurotrophic factor precursor pro-BDNF binds to TrkB and p75NTR but not to TrkA or TrkC. J Neurosci Res 80:18–28

CHAPTER 8

EVALUATION OF ANTI-PROPROTEIN CONVERTASE ACTIVITY OF DITERPENE ANDROGRAPHOLID DERIVED PRODUCTS

AJOY BASAK[1], UPENDRA K. BANIK[2], SARMISTHA BASAK, NABIL G. SEIDAH[3] AND SUIYANG LI[2]

[1] *Diseases of Aging Program, Regional Protein Chemistry Center, Ottawa Health Research Institute, Loeb Building, 725 Parkdale Ave, Ottawa, ON K1Y 4K9*
[2] *Bioscan Continental Inc., 350, Industriel Boul., St-Eustache, Quebec J7R 5R4, Canada*
[3] *Biochemical Neuroendocrinology laboratory, Clinical Research Institute of Montreal, 110 Pine Ave, W, Montréal, Que H2W 1R7, Canada*

Abstract: SEA, a mixture of succinoyl ester derivatives of andrographolid, previously shown to be the first *in vitro* nonpeptide inhibitor of Proprotein Convertases (PC) PC1, furin, and PC7 [Basak, et al., *Biochem. J.* 1999; 338: 107–113] has been tested *ex vivo* for its anti-Proprotein Convertase (PC) activity. SEA blocks PC-mediated cleavages of proBDNF (Brain Derived Neurotrophic Factor) (RVRR$^{128}\downarrow$HS) and surface envelope glycoprotein gp160 of HIV (RERR$^{511}\downarrow$AV) in a dose dependent manner. 25 μM of SEA completely blocks the cleavage of 32 kDa proBDNF into its 14 kDa mature form while 75 μM prevents significantly HIV gp160 processing into gp41. The estimated IC$_{50}$ value for the latter is $\sim$ 40 μM. At higher concentration, SEA is partly cytotoxic to the cells, with less protein secretion in culture medium. In an effort to examine further, the anti-PC activity of SEA and other andrographolid products, we examined their effects on human leukemic cell growth. Data showed a 15–50% decrease in growth in presence of SEA depending on the concentration used. These findings may provide a rationale for designing nonpeptide inhibitors of PCs based on androgholide molecule present in abundance in the medicinally active plant *Andrographis paniculata (AP)*

Keywords: Proprotein convertases, Furin, Precursor protein processing, Viral glycoproteins, Brain derived neurotrophic factor, Human Immunodeficiency Virus, gp 160, Andrographis paniculata, Lymphocyte, Protease inhibitors

A-Majid Khatib (ed.), Regulation of Carcinogenesis, Angiogenesis and Metastasis by the Proprotein Convertases, 137–154.
© 2006 *Springer.*

1.   INTRODUCTION

Proteolytic processing of larger inactive precursor proteins into smaller functionally bioactive forms is an important cellular event implicated in both normal and abnormal human pathophysiology. This hypothesis of precursor processing originally advanced more than three decades ago [1, 2], was substantiated by numerous research publications [reviewed in 3]. Because of this role, proteolytic enzymes remain in the forefront as important therapeutic targets for intervention of many diseases and metabolic disorders, even though several of these proteases are also linked to normal development and important physiological functions [3, 4]. Thus development of potent and specific protease inhibitors became an important research topic in drug design and therapeutic applications. Already a number of inhibitors directed against host as well as invading foreign proteases have been described in the literature for treatment of conditions such as viral infections like AIDS, inflammatory diseases such as chronic obstructive pulmonary disorder, prostate cancer and others [5–7]. The search for processing enzymes responsible for proteolytic activation primarily at post dibasic sites led to the discovery of Proprotein Convertases (PCs). These are a family of Ca^{2+}-dependent proteases related to bacterial subtilisin and yeast kexin. So far 7 members of PCs have been identified. They are PC1/PC3, PC2, PC4, Furin, PC4, PC5/PC6 and PC7/LPC/PC8 [reviewed in [8, 9]]. Numerous studies revealed the implication of PCs in the maturation of bacterial toxins [10, 11] and surface glycoproteins of viruses ranging from common influenza and respiratory tract infection to highly virulent ebola and HIV [12–14]. Because of this revelation, an immense interest has grown to develop specific inhibitors of PCs as potential therapeutic agents for intervention of viral and bacterial infections. Both protein based macromolecules and peptide based small molecules were described as PC inhibitor [reviewed in [15, 16]. Several studies were reported on the biochemical applications of bioengineered protein inhibitors of PCs [17–20]. However in case of small molecule PC-inhibitors, the success is very limited. For past several years our laboratory was engaged in the design of small molecule inhibitors of PCs primarily because of their enhanced metabolic and proteolytic stability and ease of accessibility. Our research led to a number of peptide inhibitors derived from the sequence of prodomain of PCs [21–25]. Lately we reported the first nonpeptide inhibitor of PCs. It is called SEA (succinoyl esters of andrographolid), a semi synthetic product derived from andrographolid the major constituent of medicinally active plant *Andrographis paniculata (AP)* [26]. SEA and its purified fractions inhibit PC1, Furin and PC7 at low micromolar concentration. Numerous studies indicated various medicinal values of AP constituents. These include (i) restoration of vasoconstriction [27], (ii) hepatoprotective activity [28], (iii) prevention of lipid peroxidation [29], (iv) antiplatelet aggregation activity [30], (v) protective effect on rat hepatocytes [31], (vi) anticholeratic activity [32], (vii) cardiovascular effect [33, 34], (viii) immunostimulant activity [35], (xi) effect on intestinal brush-border membrane-bound hydrolases [36], (x) effect on blood plasma progesterone of pregnant rats [37], (xi) antihepatotoxic activity [38], (xii) effect on hepatic microsomal drug metabolizing enzymes [39], (xiii) reduction of common cold symptoms [40], (xiv) cell differentiation inducing property [41] and most importantly, (xv) antiviral property [42, 43]. Among these, the antiviral property has drawn our particular attention because of its link to PC-activity that led to our ongoing in depth evaluation of andrographolid and its derived products.

In this article we report by *ex vivo* studies the ability of andrographolid derivative, SEA to block PC-mediated cleavages of surface glycoprotein gp160 of HIV and the neuropeptide proBDNF (Brain Derived Neurotrophic Factor) at the physiological sites (**RERR↓**) and (**RVRR↓**) respectively. In addition initial anti-proliferative activity of aqueous AP extract was also evaluated.

2.   MATERIALS AND METHODS

2.1   Cell Culture Work

Purified recombinant vaccinia virus (vv) containing full-length coding region of human pro-BDNF was used to infect cells as described previously [44]. Mouse pituitary corticotroph AtT-20 cells were cultured using Dulbecco's modified Eagle's medium (Gibco) as reported previously [45]. It was grown in 60-mm dishes and exposed to virus for 30 min at a multiplicity of 1–4 plaque-forming units (pfu) per cell. The cells were incubated in medium with or without SEA at various concentrations. Incubation was continued for another 24h. A control was done in parallel without any added SEA. The medium was harvested and BDNF was immunoprecipitated as described previously [45]. We used an affinity-purified antibody to BDNF [46], provided by Amgen Inc. at a concentration of $0.5\,\mu g/ml$. Samples were analyzed by 13–22% gradient SDS-polyacrylamide gel electrophoresis (SDS-PAGE). Gels were fixed for 1h in 40% methanol and 10% acetic acid, treated with ENHANCE (Perkin-Elmer Life Sciences) for 1 h, washed in 10% glycerol for 1 h and dried for 4 h at 60°C.

A similar protocol was used for work involving gp160 processing. The recombinant vaccinia virus of vv:gp 160 (a gift from Dr. B. Moss, NIH) and the antisera of gp160, obtained from seropositive individuals (Dr. J. Cogniaux, Institut Pasteur du Brabant) that recognizes gp160 and gp41 but not gp120 were used as described [48, 49].

2.2   Western Blot Analysis

Concentrated media from AtT20 cells expressing either BDNF or gp160 in presence or absence of SEA were applied to a 13–22% T, 2.7% C gradient SDS-PAGE and western blot analysis were performed as described in [44–48]. Blots were probed with antibodies raised against mature BDNF or gp 120. After four washes (10 min each) in Tris buffered saline containing 0.1% Tween-20, the primary antibody was visualized using horse radish peroxidase-coupled donkey antirabbit antibodies (Jackson Immuno Research Laboratories), followed by peroxidase-catalyzed chemiluminescence (enhanced chemiluminescence; Amersham) according to manufacturer's instructions.

2.3   Reverse Phase High Performance Liquid Chromatography (RP-HPLC) and Mass Spectrometry

RP-HPLC was performed using a semipreparative C_{18} Jupiter column (0.94×25cm, Phenomenex Company), followed by further purification on an analytical C_{18} Jupiter column (0.46×25cm, Phenomenex) using the conditions described earlier [21, 22]. The identity of each collected peak was confirmed by Matrix Assisted Laser Desorption Time of Flight (MALDI-TOF) mass spectrometry using Voyageur DE-Pro instrument (PE Biosystems, Framingham, Mass, USA) with either α-cyano-4-hydroxycinnamic acid (CHCA,) or 2,3-dihydroxy benzoic acid (DHB, Aldrich Chemical Co) as matrices.

2.4   Extraction of Andrographis Paniculata (AP) Whole Plant

Air dried AP plants (Bioscan Inc, Ste Eustache, Qc, Canada) (100 g) was first crushed and then extracted by reflux with ethyl acetate (3×100 ml) using a soxhlet apparatus [50]. The combined solvent was evaporated to dryness to yield a gummy residue. The residual plant material was air dried and extracted in similar manner successively with ethanol (3×100 ml), water (3×100 ml) and finally with n-butanol (3×100 ml). All extracts were separately evaporated under vacuum to yield pale yellow to brown residues. For HPLC analysis each dried residue obtained following lyophilization (5 mg) was dissolved in dimethyl formamide (DMF) (100 μl) and then diluted with 0.1% TFA/water (900 μl).

2.5   Extraction of Restorin Capsule and Restorin™

Restorin™, and AP powder (the lyophilized powder of aqueous extract of AP) were both obtained from Bioscan Continental Inc, Ste Eustache, Qc, Canada, http://www.bioscancontinental.com/index.html). The contents of two restorin™ capsules (total weight = 500mg) and AP powder (500 mg) were extracted with 0.1%TFA/water (3×10 ml) at 45 °C and lyophilized before analysis by RP-HPLC. The experiment was done in triplicates.

2.6   RP-HPLC of AP Products and its Authentic Standards

For RP-HPLC analysis, all authentic samples were dissolved in DMF (minimum volume) and then diluted with 0.1% TFA/water so as to provide a final concentration of 1 μg/μl in 5%DMF. The samples were analyzed on C_{18}- analytical column (Jupiter) using the protocol described above.

2.7   Preparation of Succinoyl Ester of Andrographolid (SEA)

SEA was prepared from andrographolid (Sigma-Aldrich Chemical Company, Milwaukee, Wisconsin, USA) by reaction with succinic anhydride as described [26]. The crude material was purified into three products (Retention times = 33.9, 36.6

and 44.0 min RP-HPLC on a semi-preparative C_{18} column [26]. The fractions SEA-33.9, SEA-36.6 and SEA-44 were characterized by MALDI-tof mass spectrometry as andrographolid $3\alpha, 14\alpha$ 19-O-trisuccinate (mol wt: observed $= 650$, calculated $= 651$), mono pyridinium andrographolid $3\alpha, 14\beta$ 19-O-trisuccinate (mol wt: observed $= 731$, calculated $= 730$) and 14-dehydro andrographolid 3α, 19-O-disuccinate (mol wt: observed $= 533$, calculated $= 532$).

2.8   Source of AP Standards

Authentic sample of andrographolid was obtained from Aldrich Chemical Co as well as from Prof. Arun K. Barua, Bose Institute, Kolkata, India while neoandrographolid, 14-deoxy andrographolid and 14 deoxy 11,12 didehydro andrographolid were gifts from Prof. J. D. Connelly, Department of Chemistry, University of Glasgow, Scotland, UK. Two additional standards of AP namely Phlogantholide A (2α-hydroxy-3-deoxyandrographolid-19-galactoside termed as **AP-1**) and phlogantholide B (2-O-galactoside of phlogantholide A, indicated as **AP-2**) were also obtained from Prof. Barua. The identity of each material was confirmed by MALDI-tof MS and thin layer chromatography (TLC) [silica gel 60 plate on aluminum sheets, 0.35mm thickness, Aldrich Chemical Company].

2.9   Effect of SEA on Cell Proliferation

Human (h)-acute T cell leukemia cell line Jurkat and the h-Burkitt's lymphoma cell line BJAB were obtained from the American Type Culture Collection (ATCC; Rockville, Maryland, USA) and maintained in RPMI 1640 (Gibco BRL) medium with 10% fetal bovine serum (Gibco BRL), 2 mM L-glutamine (Gibco BRL) and 100 units/ml penicillin-streptomycin (Gibco BRL) at 37 °C in 5% CO_2. To measure the effect of crude AP extracts on the growth of human lymphoid cells, Jurkat and BJAB cells were prepared in 100-mm plates with 10,000 cells per plate. Next day, cells were treated with 0, 25, 50 and 400 μg/ml of crude AP extracts and the viable cell number was counted after 6 days of treatment.

3.   RESULTS

3.1   RP-HPLC of AP Standards as well as Restorin™ Capsule and AP Powder Extracts

RP-HPLC chromatograms of four authentic standard components of AP namely, andrographolid (**1a**), 14-deoxy andrographolid (**2a**), neo-andrographolid (**2b**), and 14-deoxy 11,12 didehydro andrographolid (**3a**) (for complete chemical structures of these and other andrographolid derivatives see Figures 1A and 1B) were shown in Figure 2 which indicated that the samples were pure with minor impurities in some cases. Each compound appeared as a major peak with retention times (R_t) of elution as 41.9, 49.6, 52.4 and 52.9 min respectively for **3a**, **2b**, **2a** and **3a**.

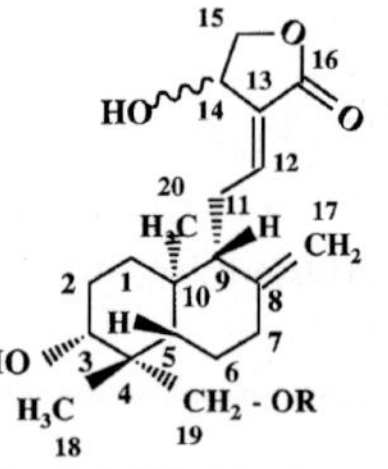

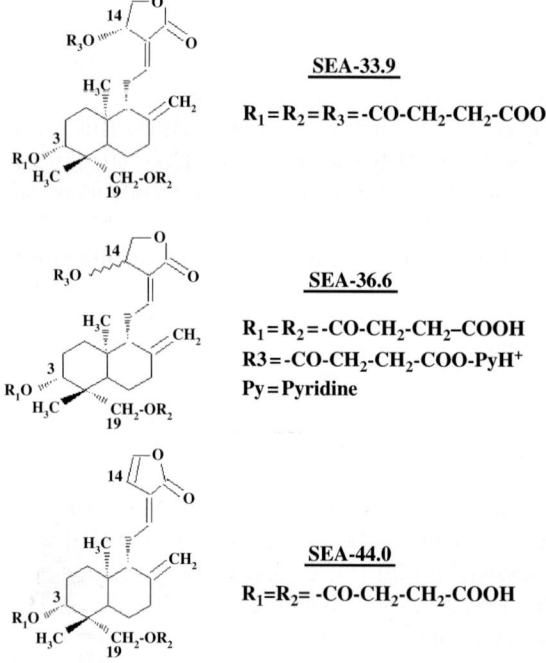

1, a = Androgapholide, 14α, R = H
b = 14-Epi andrographolide14β, R = H
c = Andrographiside, 14α, R = glc

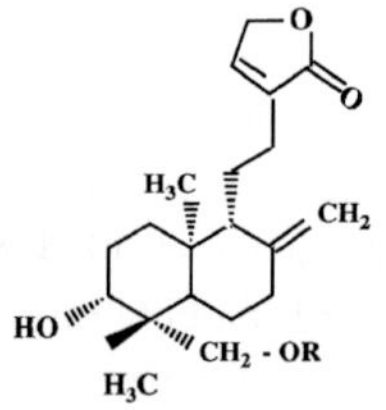

2, a = 14-Deoxy andrographolide, R = H
b = 14-Deoxy andrographiside, R = H
(Neo-andrographolide)

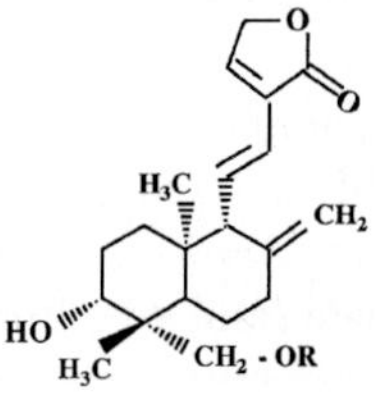

3, a = 14-Deoxy 11,12 didehydro
andrographolide, R = H
b = 14-Deoxy 11,12 didehydro
andrographiside, R = glc

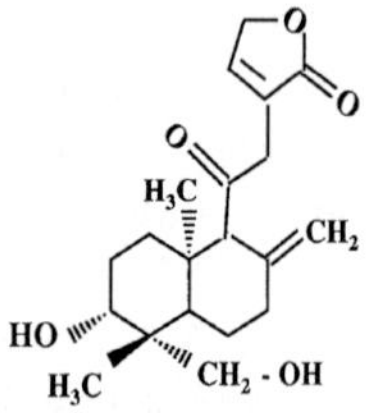

4, 14-Deoxy 11 oxoandrographolide

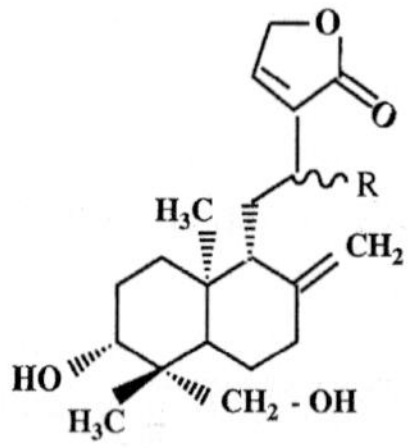

5, a = 14-Deoxy 12 methoxy andrographolide, R = β/α OMe
b = 14-Deoxy 12 hydroxy andrographolide, R = β/α O
c = 12-Epi, 14-deoxy 12 methoxy andrographolide, R = β/α OMe

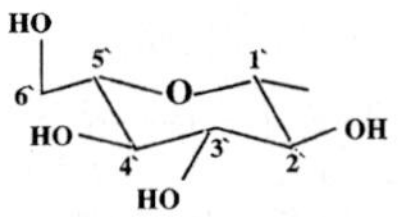

6, d(+) glucose
(glc)

Figure 1A. Chemical structures of major components of *Andrographis Paniculata*

SEA-33.9

$R_1 = R_2 = R_3 = -CO-CH_2-CH_2-COOH$

SEA-36.6

$R_1 = R_2 = -CO-CH_2-CH_2-COOH$
$R_3 = -CO-CH_2-CH_2-COO-PyH^+$
Py = Pyridine

SEA-44.0

$R_1 = R_2 = -CO-CH_2-CH_2-COOH$

Figure 1B. Chemical structures of various fractions of Succinoyl Ester of Andrographolid

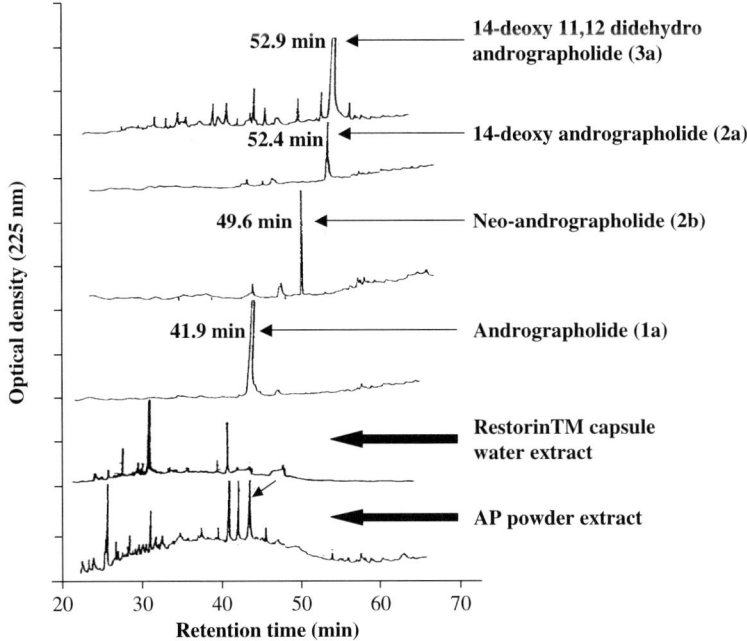

Figure 2. RP-HPLC chromatograms of authentic samples of some pure components of Andrographis Paniculata (AP) and the water extracts of Restorin TM and AP powder

The peaks were collected and confirmed by MALDI-tof mass spectrometry. When compared with the RP-HPLC chromatograms of Restorin™ capsule and AP powder aqueous extracts, it is evident that both extracts contain andrographolid (the major component of AP) as one of the constituents but the amount is more significant in AP powder compared to Restorin™ capsule.

Figure 3 shows the RP-HPLC chromatograms of various crude extracts (both non-aqueous and aqueous) of AP. Clearly depending on the nature of the extract each contains a large variety of components including andrographolid and neoandrographolid as indicated in the Figure.

Based on a standard curve with authentic andrographolid (*not shown*), we estimated the presence of 0.05 % and 0.6% of andrographolid in Restorin™ and AP powder respectively. Both contain a small amount of neoandrographolid (~0.01%) while none of them seem to contain any significant amounts of either 14-deoxy compounds **2a** and **3a**. In addition both AP-capsule (Restorin™) and AP-powder contain a variety of other chemical compounds.

3.2   Effects of SEA on Cleavage of proBDNF by Endogenous Furin

Since many of the natural constituents of AP and one of its chemically modified products, SEA were shown to inhibit proprotein convertase activity including furin

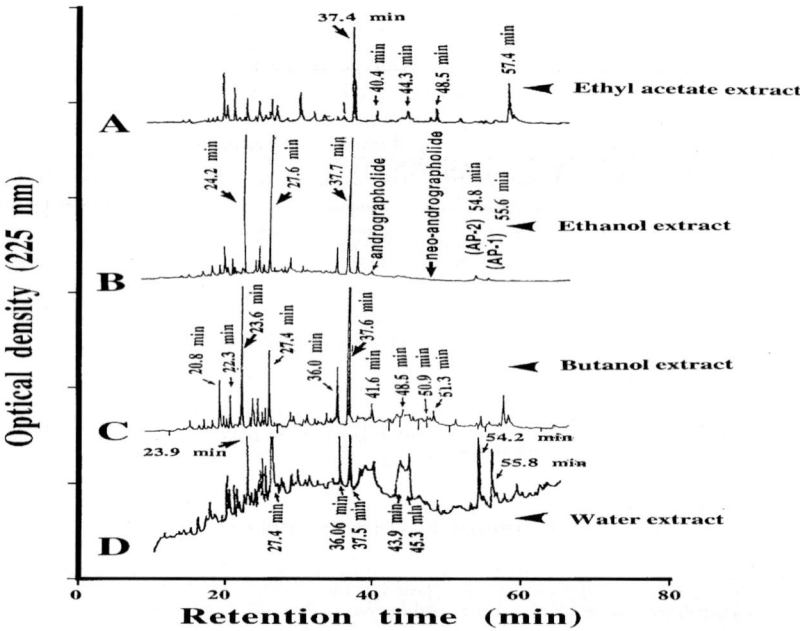

Figure 3. RP-HPLC chromatograms of various crude aqueous and non-aqueous extracts of AP

in vitro [26], we decided to examine *ex vivo* the anti-PC activity of SEA, the most potent andrographolide based PC-inhibitor known so far.

As a model target, we used the furin dependent processing of proBDNF into mature BDNF in AtT20 cells and monitored the processing on western blot using BDNF antibody. The initially produced 32 kDa inactive proBDNF is processed by the proprotein convertase furin at **RVRR**[128]**↓HS** site to generate the functionally active 14 kDa BDNF. The sequence of cleavage events is shown schematically in Figure 4A. Thus furin activity is considered as one of the most crucial steps in the maturation of this important neurotropic factor [45].

Effect of SEA on this cleavage when added exogenously to the medium is shown in Figure 5. Both crude SEA containing mostly tri-succinoyl ester of andrographolid as well its three purified fractions, **SEA-33.9, 36.6** and **44.0** (for structures see Figure 1B) were used for this study. As shown in Figure 5, both panels, it is evident that in the control (no SEA added), proBDNF (32kDa) is completely cleaved as expected by endogenous proprotein convertase furin into its 14 kDa mature form (extreme left lane). However presence of SEA at 15 μg/ml concentration is able to block this cleavage to a significant extent while a 10-fold lower concentration (0.5 μg/ml) is unable to do so (left panel). When purified SEA fractions were tested under similar condition except at 10 μg/ml concentration, only **SEA-44.0** (14-deoxy andrographolid 3α, 19, O-disuccinate) is capable of suppressing this cleavage, as indicated by the appearance of the unprocessed 32kDa band of proBDNF.

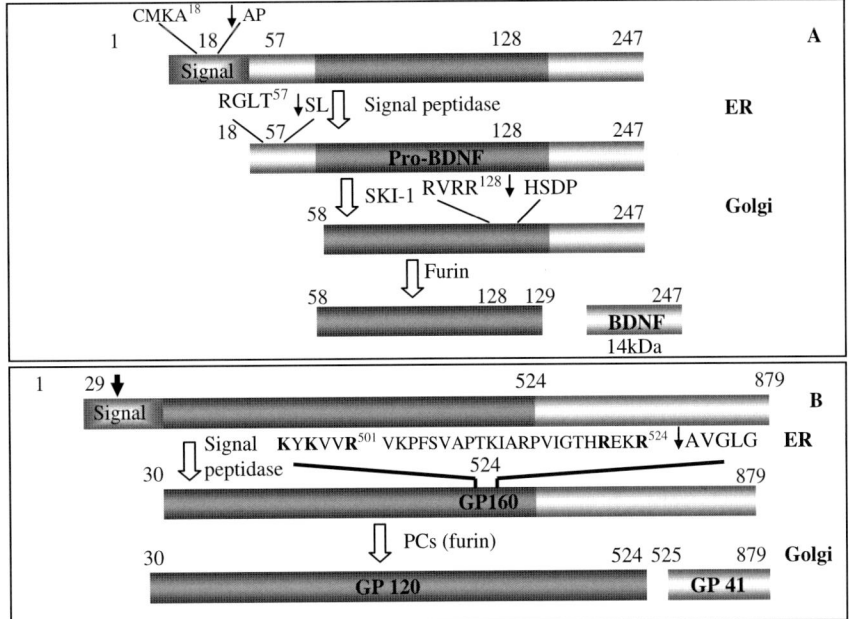

Figure 4. Schematic representation of steps involved in the maturation of proBDNF (A) and gp160 of HIV (B)

Neither **SEA-36.6** (monopyridinium andrographolid 3α, 14α 19-trisuccinate) nor **SEA-33.9** (3α, 14α, 19-trisuccinate) is able to block this cleavage at the same concentration level.

3.3 Dose-dependent Effect of SEA-44.0 on proBDNF Cleavage by Furin

Anti-proteolytic activity of SEA-44.0 towards the furin-mediated cleavage of proBDNF to BDNF at various concentrations ranging from 10–20 μg/ml is shown in Figure 6 (left panel). It revealed that there is a gradual increase in the amount of unprocessed proBDNF produced as the concentration of added SEA-44.0 is increased from 10 to 20 μg/ml. The data is also presented in densitometric bar graph format (Figure 6, right panel) where it is noted that the ratio of proBDNF:BDNF increases with the amount of SEA-44.0 used. It may be pointed out that in the control lane with no added SEA-44.0, the amount of proBDNF band is minimal as expected. At concentration > 50 μg/ml, secretion of either mature BDNF or its precursor form into the medium is blocked significantly (*data not shown*).

3.4 Effect of SEA on PC-mediated Cleavage of gp160 of HIV

In order to confirm further the anti-convertase activity of SEA *ex vivo*, another convertase substrate gp160, the surface viral glycoprotein of HIV-1 virus was tested

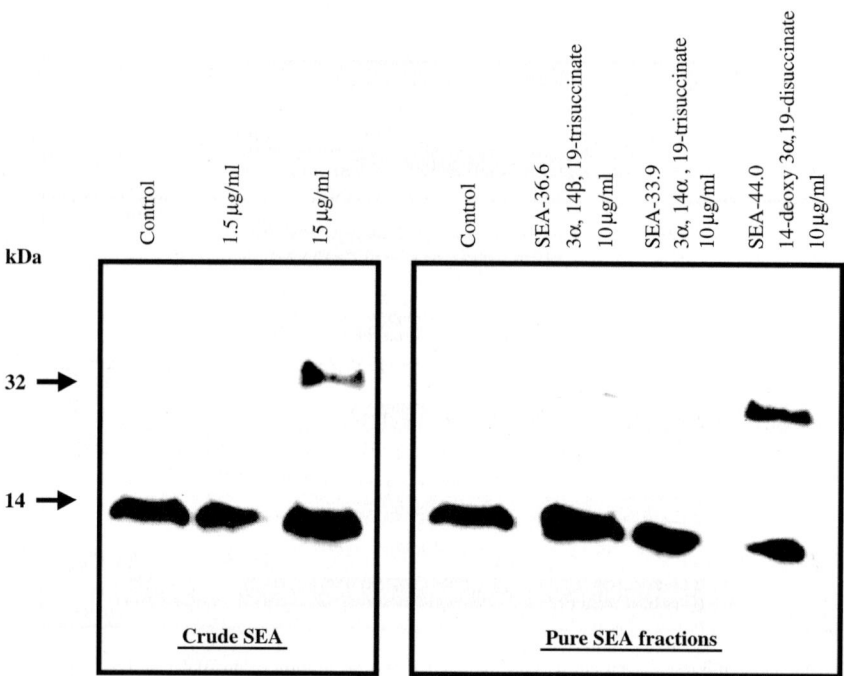

Figure 5. Effects of crude and purified fractions of SEA on furin-mediated processing of h-proBDNF, using BDNF antibody western blot analysis

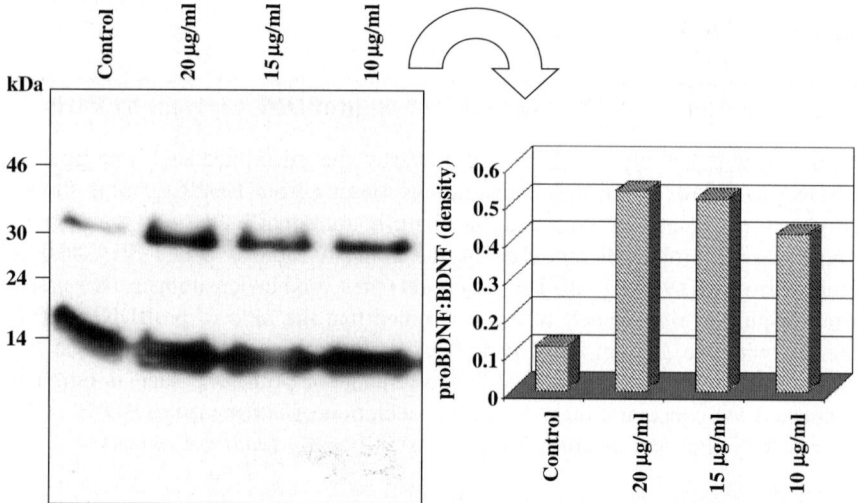

Figure 6. Dose dependent effects of SEA-44.5 (andrographolid 14-deoxy 3a,19-disuccinate) on furin-mediated processing of h-proBDNF. Left: 12% SDS-PAGE gel picture western blot using BDNF antibody, Right: bar graph representation showing the ratio of proBDNF:BDNF at various condition

in AtT20 cells. PC-mediated cleavage of gp160 into gp120 and gp41 was monitored in the absence and presence of various concentrations of added SEA and PDX- a known inhibitor of PCs including furin (gift from G. Thomas, Vollum Institute, Portland, USA). The western blot analysis results using anti-gp 160 antibody which also recognizes the cleaved gp 41 but not gp120 fragment (for cleavage scheme see Figure 1, panel B) is shown in Figure 7.

As expected the control lane with no added SEA, showed a much increased level of processed gp41 protein. The data indicated that crude SEA blocks gp160 processing to gp 41 in a dose dependent manner. There is a gradual decrease in the amount of gp 41 produced as the concentration of added crude SEA is increased. In our condition, 10–25 µg/ml of crude SEA is as effective as 1–5 µg/ml PDX in blocking gp 160 processing. As observed with proBDNF processing, a higher dose of SEA (> 50 µM) also prevented secretion of both gp 41 and its precursor gp 160 into the culture medium. In contrast to proBDNF experiment, here we used only crude SEA instead of its purified fractions since initial data indicated no significant differences on their effects on gp160 processing.

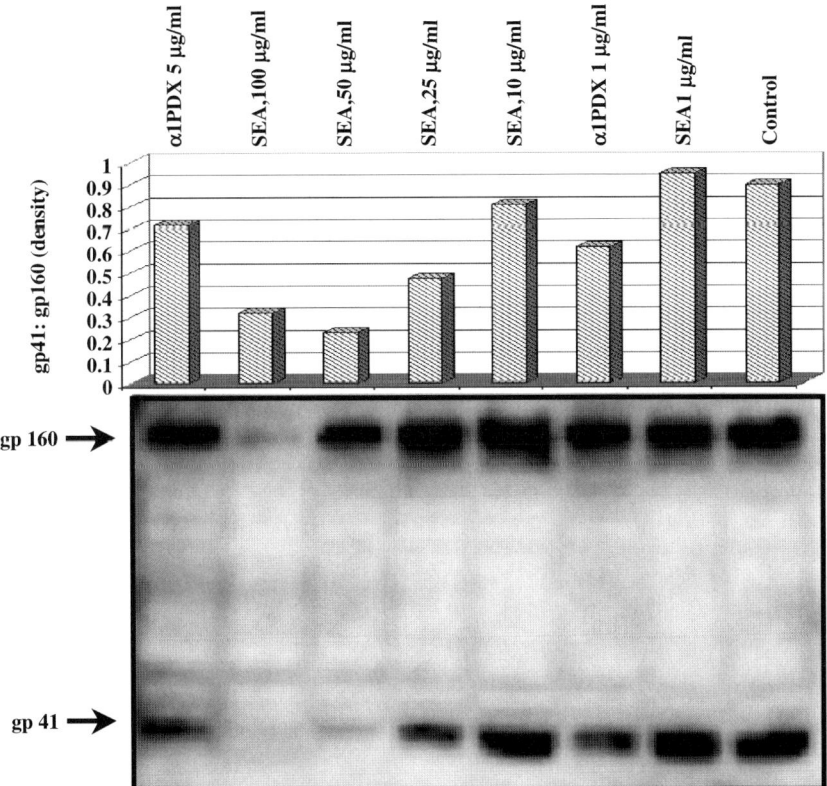

Figure 7. Effect of crude SEA on furin-mediated processing of glycoprotein gp160 of HIV

3.5   Effect of Andrographolide Extract on Cell Proliferation

Since proprotein convertases (PCs) particularly furin has been linked to the maturation of a large number of growth factors involved in cell proliferation event, we decided to test first potential antiproliferation activity of crude AP extracts which exhibit modesrate PC-inhibitory activity (unpublished). We used Jurkat and BJAB cell lines for the study. Jurkat is a T cell leukemic cell line, whereas BJAB is a B cell leukemic cell line. AP extracts inhibited both T cell and B cell leukemia cell growth at higher concentrations, suggesting a broad and moderate anti-proliferative effect of AP extracts (Table 1 and Figure 8). The result was the average of three independent experiments. This effect is particularly evident for

Table 1. Mean viable cell number upon treatment with AP extract. Cell number was counted after 6 days of treatment. Promotive/cytotoxic effects of AP extract on lymphocyte growth were expressed in percentages different from control

Concentration (µg/ml)	Jurkat (T lymphocyte)		BJAB (B lymphocyte)	
	(×1,000 cells)[a]	(%)[b]	(×1,000 cells)	(%)
0	961	N/A	528	N/A
25	1079	+12.3	545	+3.2
50	1128	+17.4	484	−8.3
400	845	−12.1	243	−54.0

[a] number of cells.
[b] percentage of cells.

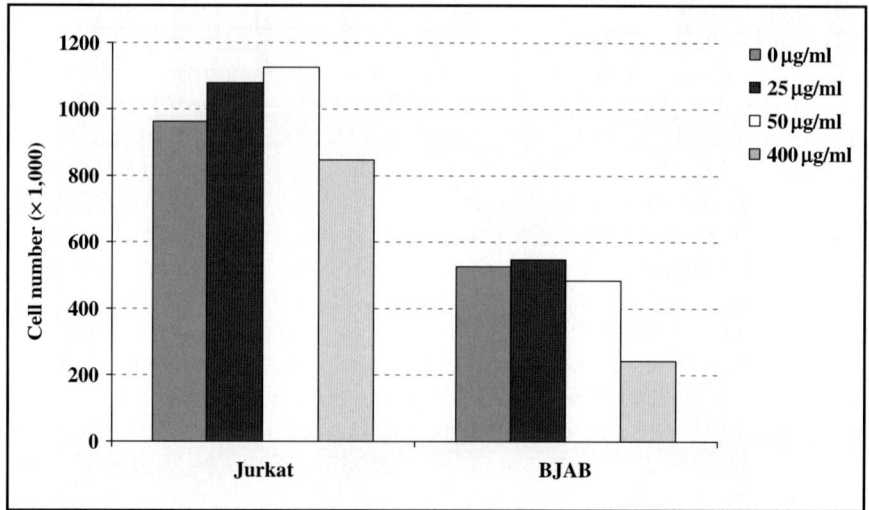

Figure 8. Effect of crude aqueous AP extract on lymphocyte growth

BJAB cell growth where a dose dependent effect was observed. For the other cell line an initial increase in lymphocyte growth is followed by a subsequent decrease. At relatively high concentration of $400\,\mu g/ml$ of added AP-extract, a 15% and 50% decreased lymphocyte growth was observed for Jurkat and BJAB cell lines respectively (Figure 8). Again the data is the average of three independent experiments.

4. DISCUSSION

SEA, the chemically modified form of andrographolid blocks endogenous furin/PC dependent cleavages of both proBDNF and the viral glycoprotein gp160 of HIV in a dose-dependent manner. The proteolytic maturation of these precursor proteins has been the subject of intense research of many studies. This is primarily because of the implication of mature BDNF in survival, growth and maintenance of neurons in central and peripheral nervous system [44, 45], while gp160 processing is required for viral pathogenesis of HIV [48, 49]. Intracellular cleavage of proneurotrophins such as proBDNF to produce active growth factors like BDNF occurs following pairs of basic amino acids and is mostly mediated by furin, PC5 and/or PC7. Previously furin inhibitor PDX has been shown to block this cleavage and hence cell growth and proliferation indicating their possible role as anti-proliferative agents in cancer therapy [44, 45]. Present study revealed that like PDX, the nonpeptide PC-inhibitor SEA consisting of a mixture of di and tri succinoyl esters of andrographolid is also able to block the maturation of proBDNF to BDNF, with less efficiently. Using pure fractions of SEA, it is revealed that this bioactivity is mainly due to SEA 44.0 (andrographolid 14-deoxy 3α,19 O-disuccinate) which has already been shown to inhibit *in vitro* PC-activity against small peptide fluorogenic substrates [26].

The data further indicated that crude SEA blocks the processing of gp 160 into gp 41 in a dose-dependent manner. Since this processing is mediated by PCs such as furin, it is very likely that again SEA exhibits this activity because of its inhibitory role on PC-activity. Unlike proBDNF this bioactivity is not linked to any particular SEA fraction but the entire crude SEA. In consistent with previous *in vitro* study, this *ex vivo* data suggests that SEA is an inhibitor of PCs [26]. Furthermore many of the biological properties of AP plant and its products including its antiviral and immune stimulatory effects may be linked to this unique biological property which may be enhanced following appropriate chemical modification of andrographolide. A recent clinical study by us on a single human HIV-positive subject, revealed a strong antiviral property of Restorin™ as when taken orally with other combination drugs [51]. This may suggest an alternative approach for intervention of HIV virulence where both the invading viral coat proteases such as reverse transcriptase, integrase and HIV protease and the host protease such as PCs implicated in gp160 maturation, are both targeted at the same time. Such a host-guest derived combination approach may help in bringing the viral load and pathogenic condition under control so that the patients can live longer with the disease. Such anapproach is more logical in light of the previous study of Chang and Yeung who reported an inhibition of growth of HIV (by 97–100%) *in vitro* by crude hot water extract of AP [52]. It is believed that SEA interfered with the growth of HIV by a mechanism other than interferon induction. Studies indicated that it (i) did not inactivate cellular HIV, (ii) inhibit reverse trancriptase activity (iii) only partially interfered with the HIV-induced cell fusion and with binding of HIV to the T helper cell line and in phytohemagglutinin activated blood mononuclear cells from HIV seronegetive persons and finally (iv) partially inhibited HIV replication after the virus had attached to cell receptors [51]. Thus SEA exerts its suppressive effect on HIV growth by interference with the binding of virions to cells and with a step in the viral replication cycle subsequent to virus-cell binding. Its effect on the processing of gp160 will now provide a possible answer to its anti-HIV property.

Another important aspect of this study is that at concentration $> 50\,\mu g/ml$, SEA prevents the secretion of BDNF and gp 41 as well as their precursor proteins into the culture medium of the cell line used. The mechanism for this observation is not clearly understood. However it may suggest an alternative pathway for regulation of these bioactive peptides in circulation by SEA and AP products.

Our observation that AP aqueous extract possesses some antiproliferative activity is consistent with similar and more extensive studies done with several PC inhibitors including PDX [18, 54]. Even though the effect is modest with AP-aqueous extract, the data provides rationale for observed anti-tumorgenic activity of AP plant and some of its extracts [55, 57]. In aother article Lan et al. [58] has shown the growth inhibitory activity of andrographolid and 14-dcoxy-11,12-didchydroandrographolid against a number of human tumor cell lines. These compounds showed more activity in terms of EC_{50} as compared to vincristine sulphate and etoposide, especially against T-47D cells, suggesting them as possible candidates against breast cancer. Additional studies are needed to confirm these results and also to identify the active principles of AP.

5.   CONCLUSION

In conclusion, this study has revealed that the diterpene andrographolid derived compound SEA exhibits *ex vivo* anti-PC activity towards the proteolytic maturation of proBDNF and gp160 of HIV-1. SEA was previously shown *in vitro* to inhibit the enzymatic activity of Proprotein Convertases PC1, furin and PC7 against small fluorogenic peptide substrates [26]. Taken together, these findings and the previously reported case study [51] suggested that various observed bioactivities of AP plant constituents and its extracts may be related to their (or their chemically modified compounds) anti-PC activities and that such products are capable of exhibiting their activities *in vitro* and *ex vivo* models. The ability of AP aqueous extract to inhibit lymphocyte cell growth at least to a modest extent and promotion is highly significant because of its potential role in tumorgenesis and cancer. Further studies are required to substantiate this claim.

ACKNOWLEDGEMENTS

We thank Dr. Michel Chrétien, Diseases of Aging, OHRI, Ottawa, Canada for support and encouragement during the initial phase of this work. We thank Dr. Sam Cooper, formerly INRS-Santé, UQAM, Montreal, Dr. Bakary B. Touré, Clinical Research Institute and Dayani Mohottalage, OHRI for conducting extractions, initial experiments, and technical assistance. This research is funded by UOttawa, Bioscan and Canadian Foundation of Innovation new opportunity funds.

Abbreviations: PC, Proprotein Convertase; HPLC, high performance liquid chromatography; BDNF, brain derived neurotriphic factor; gp160, glycoprotein 160; HIV, human immunodeficiency virus; SEA, Succinoyl ester of andrograppholid crude; SEA-33.9, SEA with HPLC retention time 33.9 min; SEA-36.6, SEA with HPLC retention time 36.6 min; SEA-44.0, SEA fraction with retention time 44.0 min; AP, Andrographis Paniculata; AT, antitrypsin, PDX, antitrypsin Portland; tris, tris -hydroxymethyl)-aminomethane; MALDI-tof MS, matrix assisted laser desorption time of flight mass spectroscopy.

REFERENCES

[1] Chrétien M, Li CH (1967) Isolation, purification and characterization of c lipotropic hormpone from sheep pituitary glands. Can. J. Biochem. **45**:1163–1174

[2] Steiner DF, Cunningham D, Spiegelman L, Aten B (1967) Insulin biosynthesis: Evidence for a precursor. Science. **157**:697–699

[3] Deadman J (2000) Proteinase inhibitors and activators strategic targets for therapeutic intervention. J. Pept. Sci. **6**:421 431

[4] Southan C (2000) Assessing the protease and protease inhibitor content of the human genome. "Proteinase inhibitors and activators: Strategic targets for therapeutic intervention", University of Oxford, 17–20

[5] Behrens G, Dejam A, Schmidt H, Balks HJ, Brabant G, Korner T, Stoll M, Schmidt RE (1999) Impaired glucose tolerance, beta cell function and lipid metabolism in HIV patients under treatment with protease inhibitors. AIDS. **13**:F63–F70. (Also see: http://www.hhs.gov/news/press/2004pres/20040516a.html and http://www.thebody.com/treat/newdrugs.html.

[6] Mikolajczyk1 SD, Millar LS, Marker KM, Rittenhouse HG, Wolfert RL, Marks LS, Charlesworth MC, Tindall DJ (1999) Identification of a Novel Complex between Human Kallikrein 2 and Protease Inhibitor-6 in Prostate Cancer Tissue. Cancer Research **59**:3927–3930

[7] Song XY, Zeng L, Jin W, Thompson J, Mizel DE, Lei KJ, Billinghurst RC, Poole RA, Wahl SM (1999) Secretory Leukocyte Protease Inhibitor Suppresses the Inflammation and Joint Damage of Bacterial Cell Wall–induced Arthritis. Exp Med **190**:535–542. (Also see http://www.arrivapharm.com/product-copd.php)

[8] Seidah NG, Chrétien M (1999) Proprotein and prohprmone convertases: A family of subtilases generating diverse bioactive polypeptides. Brain Research **848**:45–62

[9] Steiner DF (1998) The proprotein Convertases. Curr Opin Chem Biol **2**:31–39

[10] Thomas G (2002) Furin at the cutting edge: from protein traffic to embryogenesis and disease. Nat Rev Mol Cell Biol **3**:753–766

[11] Sarac MS, Peinado JR, Leppla SH, Lindberg I (2004) Protection against anthrax toxemia by hexa-D-arginine in vitro and in vivo. Infect Immun. **72**:602–605

[12] Jean F, Thomas L, Molloy SS, Liu G, Jarvis MA, Nelson JA, Thomas G (2000) A protein-based therapeutic for human cytomegalovirus infection. Proc Nat Acad Sci USA **97**:2864–2869

[13] Feldmann H, Volchkov VE, Volchkova VA, Klenk HD (1999) The glycoproteins of Marburg and Ebola virus and their potential roles in pathogenesis. Arch Virol Suppl 15:159–169

[14] Zambon MC (2001) The pathogenesis of influenza in humans. Rev Med Virol 11:227–241

[15] Fugere M, Day R (2002) Inhibitors of the subtilase-like pro-protein convertases (SPCs). Curr Pharm Des 8:549–562

[16] Basak, A (2005) Proprotein Convertase inhibitors: Minireview. J. Mol. Med. 83:844–855

[17] Bassi DE, Mahloogi H, Lopez De Cicco R, Klein-Szanto A (2003) Increased furin activity enhances the malignant phenotype of human head and neck cancer cells. Am J Pathol 62:439–447

[18] Khatib AM, Siegfried G, Prat A, Luis J, Chretien M, Metrakos P, Seidah NG (2001) Inhibition of proprotein convertases is associated with loss of growth and tumorigenicity of HT-29 human colon carcinoma cells: Importance of insulin-like growth factor-1 (IGF-1) receptor processing in IGF-1-mediated functions. J Biol Chem 276:30686–30693

[19] Mowla SJ, Farhadi HF, Pareek S, Atwal JK, Morris SJ, Seidah NG, Murphy RA (2001) Biosynthesis and post-translational processing of the precursor to brain-derived neurotrophic factor. J Biol Chem 276:12660–12666

[20] Tsuji A, Hashimoto E, Ikoma T, Taniguchi T, Mori K, Nagahama M, Matsuda Y (1999) Inactivation of proprotein convertase, PACE4, by alpha1-antitrypsin Portland (alpha1-PDX), a blocker of proteolytic activation of bone morphogenetic protein during embryogenesis: Evidence that PACE4 is able to form an SDS-stable acyl intermediate with alpha1-PDX. J Biochem (Tokyo). 126:591–603

[21] Jean F, Basak A, DiMaio J, Seidah NG, Lazure C (1995) An internally quenched fluorogenic PC1 and furin substrate leads to a potent prohormone convertase inhibitor. Biochem J 307:689–695

[22] Basak A, Schmidt C, Ismail AA, Seidah NG, Chrétien M, Lazure C (1995) Peptidyl Substrates containing Unnatural amino acid at the P′1 positio are Potent Inhibitors of Prohormone Convertases. Int J Pept Prot Res 46:228–237

[23] Jean F, Boudreault A, Basak A, Seidah NG, Lazure C (1995) Fluorescent peptidyl substrates as an aid in studying the substrate specificity of human prohormone convertase PC1 and furin and in designing a potent irreversible inhibitor. J Biol Chem 270:19225–19231

[24] Zhong M, Munzer JS, Basak A, Benjannet S, Mowla SJ, Decroly E, Chrétien M, Seidah NG (1999) The prosegments of Furin and PC7 as potent inhibitors of proprotein convertases: In vitro and ex vivo assessment of their specificity and selectivity. J Biol Chem 274:33913–33920

[25] Basak A, Lazure C (2003) Synthetic peptides of prosegments of proprotein convertases PC1 and furin are inhibitors of both convertases with some selectivity. Biochem J 373:231–239

[26] Basak A, Cooper S, Roberge AG, Banik UK, Chrétien M, Seidah NG (1999) Inhibition of proprotein convertases-1, -7 and furin by diterpines of Andrographis paniculata and their succinoyl esters. Biochem J 338:107–113

[27] Chiou WF, Lin JJ, Chen CF (1998) Andrographolid suppresses the expression of inducible nitric oxide synthase in macrophage and restores the vasoconstriction in rat aorta treated with lipopolysaccharide. Br J Pharmacol 125:327–334

[28] Handa SS, Sharma A (1990) Hepatoprotective activity of andrographolid from Andrographis paniculata against carbontetrachloride. Indian J Med Res 92:276–283

[29] Choudhury BR, Poddar MK (1984) Andrographolid and kalmegh (Andrographis paniculata) extract: in vivo and in vitro effect on hepatic lipid peroxidation. Methods Find Exp Clin Pharmacol 6:481–485

[30] Amroyan E, Gabrielian E, Panossian A, Wikman G, Wagner H (1999) Inhibitory effect of andrographolid from Andrographis paniculata on PAF-induced platelet aggregation. Phytomedicine, 6:27–31

[31] Visen PK, Shukla B, Patnaik GK, Dhawan BN (1993) Andrographolid protects rat hepatocytes against paracetamol-induced damage. J Ethnopharmacol 40:131–136

[32] Shukla B, Visen PK, Patnaik GK, Dhawan BN (1992) Choleretic effect of andrographolid in rats and guinea pigs. Planta Med 58:146–149

[33] Aminuddin A, Girach RD (1991) (Surv Medicinal Plants Unit, Regl Res Inst Unani Med, Bhadrak 756100). Ethnobotanical studies on Bondo tribe of district Koraput (Orissa), India. Ethnobotany **3** (1&2):15–19

[34] Zhang CY, Tan BK (1997) Mechanisms of cardiovascular activity of Andrographis paniculata in the anaesthetized rat. J Ethnopharmacol **56**:97–101

[35] Puri A, Saxena R, Saxena RP, Saxena KC, Srivastava V, Tandon JS (1993) Immunostimulant agents from Andrographis paniculata. J Nat Product **56**:995–999

[36] Choudhury BR, Poddar MK (1985) Andrographolid and kalmegh (Andrographis paniculata) extract: Effect on intestinal brush-border membrane-bound hydrolases. Methods Find Exp Clin Pharmacol **7**:617–621

[37] Panossian A, Kochikian A, Gabrielian E, Muradian R, Stepanian H, Arsenian F, Wagner H (1999) Effect of Andrographis paniculata extract on progesterone in blood plasma of pregnant rats. Phytomedicine **6**:157–161

[38] Kapil A, Koul IB, Banerjee SK, Gupta BD (1993) Antihepatotoxic effects of major diterpenoid constituents of Andrographis paniculata. Biochem Pharmacol **46**:182–185

[39] Choudhury BR, Haque SJ, Poddar MK (1987) In vivo and in vitro effects of kalmegh (Andrographis paniculata) extract and andrographolid on hepatic microsomal drug metabolizing enzymes. Planta Med. **53**:135–140

[40] Cáceres DD, Hancke JL, Burgos RA, Sandberg F, Wikman GK (1999) Use of visual analogue scale measurements (VAS) to assess the effectiveness of standardized Andrographis paniculata extract SHA-10 in reducing the symptoms of common cold. A randomized double blind-placebo study. Phytomedicine **6**:217–223

[41] Matsuda T, Kuroyanagi M, Sugiyama S, Umehara K, Ueno A, Nishi K (1994) Cell differentiation-inducing diterpenes from *Androgaphis paniculata Nees*. Chem Pharm Bull (Tokyo). **42**:1216–1225

[42] Mehrotra R, Rawat S, Kulshreshtha DK, Patnaik GK, Dhawan BN (1990) In vitro studies on the effect of certain natural products against hepatitis B virus. Indian J Med Res **92**:133–138

[43] Chang RS, Ding L, Chen GQ, Pan QC, Zhao ZL, Smith KM (1991) Dehydroandrographolid succinic acid monoester as an inhibitor against the human immunodeficiency virus. Proc Soc Exp Biol Med **197**:59–66

[44] Seidah NG, Benjannet S, Pareek S, Chretien M, Murphy RA (1996) Cellular processing of the neurotrophin precursors of NT3 and BDNF by the mammalian proprotein convertases. FEBS Lett **379**:248–250

[45] Seidah NG, Benjannet S, Pareek S, Savaria D, Hamelin J, Goulet B, Lalibete J, Lazure C, Chretien M, Murphy RA (1996) Cellular processing of the nerve growth factor precursor by the mammalian pro-protein convertases. Biochem J **314**:951–960

[46] Mowla SJ, Pareek S, Farhadi HF, Petrecca K, Fawcett JP, Seidah NG, Morris SJ, Sossin WS, Murphy RA (1999) Differential sorting of nerve growth factor and brain-derived neurotrophic factor in hippocampal neurons. J Neurosci **19**:2069–2080

[47] Yan Q, Rosenfeld RD, Matheson CR, Hawkins N, Lopez OT, Bennett L, Welcher AA (1997) Expression of brain-derived neurotrophic factor protein in the adult rat central nervous system. Neuroscience **78**:431–448

[48] Wouters S, Decroly E, Vandenbranden M, Shober D, Fuchs R, Morel V, Leruth M, Seidah NG, Courtoy PJ, Ruysschaert JM (1999) Occurrence of an HIV-1 gp160 endoproteolytic activity in low-density vesicles and evidence for a distinct density distribution from endogenously expressed furin and PC7/LPC convertases. FEBS Lett **456**:97–102

[49] Vollenweider F, Benjannet S, Decroly E, Savaria D, Lazure C, Thomas G, Chretien M, Seidah NG (1996) Comparative cellular processing of the human immunodeficiency virus (HIV-1) envelope glycoprotein gp160 by the mammalian subtilisin/kexin-like convertases. Biochem J **314**:521–532

[50] Pedersen JR, Olsson JO (2003) Soxhlet extraction of acrylamide from potato chips. Analyst **128**:332–334

[51] Basak A, Zhong M, Munzer JS, Chretien M, Seidah NG (2001) Implication of the proprotein convertases furin, PC5 and PC7 in the cleavage of surface glycoproteins of Hong Kong, Ebola

and respiratory syncytial viruses: A comparative analysis with fluorogenic peptides. Biochem J **353**:537–545

[52] Basak A, Li S, Banik UK (2003) A new combination drugs using andrographolid derived natural product Restomune for management of HIV: A case study. Case Reports and Clinical Practice Review **4**:223–233

[53] Chang RS, Yeung HW (1988) Inhibition of growth of human immunodeficiency virus in vitro by crude extracts of Chinese medicinal herbs. Antiviral Research **9**:163–176 (http://www.aids.org/atn/a-061-04.html)

[54] Bassi DE, Lopez De Cicco R, Mahloogi H, Zucker S, Thomas G, Klein-Szanto AJ (2001) Furin inhibition results in absent or decreased invasiveness and tumorigenicity of human cancer cells. Proc Natl Acad Sci USA **98**:10326–10331

[55] (http://www.geocities.com/andrographis/AndroBaseText.txt)

[56] Rajagopal S, Kumar RA, Deevi DS, Satyanarayana C, Rajagopalan R (2003) Andrographolide, a potential cancer therapeutic agent isolated from Andrographis paniculata. J Exp Ther Oncol **3**:147–158

[57] Matsuda T, Kuroyanagi M, Sugiyama S, Umehara K, Ueno A, Nishi K (1994) Cell differentiation-inducing diterpenes from Andrographis paniculata Nees. Chem Pharm Bull (Tokyo) **42**:1216–1225

[58] Lan MT, Sifzizul Muhammad TST, Kuroyanagi M, Sulaiman SF, Najimuddn N: Determination of growth inhibitory activities of several bioactive compounds of Andrographis paniculata a panel of human tumor cell lines: 14-deoxy-11, 12- didehydroandrographolid induces a non-apoptotic programmed cell death in T-47D, a breast carcinoma cell line. www.usm.my/nsf/sample%20full%20paper3-20Sept.pdf.

INDEX